ASE Test Preparation

Automotive Technician Certification Series

Suspension and Steering (A4)

5th Edition

❖ Cengage

Australia • Brazil • Canada • Mexico • Singapore • United Kingdom • United States

ASE Test Preparation: Automotive Technician Certification Series, Suspension and Steering (A4), 5th Edition

Vice President, Technology and Trades Professional Business Unit: Gregory L. Clayton

Director, Professional Transportation Industry Training Solutions: Kristen L. Davis

Product Manager: Katie McGuire

Editorial Assistant: Danielle Filippone

Director of Marketing: Beth A. Lutz

Marketing Manager: Jennifer Barbic

Senior Production Director: Wendy Troeger

Production Manager: Sherondra Thedford

Content Project Manager: PreMediaGlobal

Senior Art Director: Benj Gleeksman

Section Opener Image: Image Copyright Creations, 2012. Used under License from Shutterstock.com

ISBN-13: 978-1-111-12706-0
ISBN-10: 1-111-12706-9

Cengage
200 Pier 4 Boulevard
Boston, MA 02210
USA

Cengage is a leading provider of customized learning solutions with employees residing in nearly 40 different countries and sales in more than 125 countries around the world. Find your local representative at: **www.cengage.com.**

For more information on transportation titles available from Cengage Learning, please visit our website at **www.trainingbay.cengage.com.**

To learn more about Cengage platforms and services, register or access your online learning solution, or purchase materials for your course, visit **www.cengage.com.**

Table of Contents

SECTION 6   Answer Keys and Explanations 105

SECTION 7   Appendices 205

Preface

Delmar, a part of Cengage Learning, is very pleased that you have chosen to use our ASE Test Preparation Guide to help prepare yourself for the Suspension and Steering (A4) ASE certification examination. This guide is designed to help prepare you for your actual exam by providing you with an overview and introduction of the testing process, introducing you to the task list for the Suspension and Steering (A4) certification exam, giving you an understanding of what knowledge and skills you are expected to have in order to successfully perform the duties associated with each task area, and providing you with several preparation exams designed to emulate the live exam content in hopes of assessing your overall exam readiness.

If you have a basic working knowledge of the discipline you are testing for, you will find this book is an excellent guide, helping you understand the "must know" items needed to successfully pass the ASE certification exam. This manual is not a textbook. Its objective is to prepare the individual who has the existing requisite experience and knowledge to attempt the challenge of the ASE certification process. This guide cannot replace the hands-on experience and theoretical knowledge required by ASE to master the vehicle repair technology associated with this exam. If you are unable to understand more than a few of the preparation questions and their corresponding explanations in this book, it could be that you require either more shop-floor experience or further study.

This book begins by providing an overview of, and introduction to, the testing process. This section outlines what we recommend you do to prepare, what to expect on the actual test day, and overall methodologies for your success. This section is followed by a detailed overview of the ASE task list to include explanations of the knowledge and skills you must possess to successfully answer questions related to each particular task. After the task list, we provide six sample preparation exams for you to use as a means of evaluating areas of understanding, as well as areas requiring improvement in order to successfully pass the ASE exam. Delmar is the first and only test preparation organization to provide so many unique preparation exams. We enhanced our guides to include this support as a means of providing you with the best preparation product available. Section 6 of this guide includes the answer keys for each preparation exam, along with the answer explanations for each question. Each answer explanation also contains a reference back to the related task or tasks that it assesses. This will provide you with a quick and easy method for referring back to the task list whenever needed. The last section of this book contains blank answer sheet forms you can use as you attempt each preparation exam, along with a glossary of terms.

OUR COMMITMENT TO EXCELLENCE

Thank you for choosing Delmar, Cengage Learning for your ASE test preparation needs. All of the writers, editors, and Delmar staff have worked very hard to make this test preparation guide second to none. We feel confident that you will find this guide easy to use and extremely beneficial as you prepare for your actual ASE exam.

Delmar, Cengage Learning has sought out the best subject matter experts in the country to help with the development of *ASE Test Preparation: Automotive Technician Certification Series, Suspension and Steering (A4), 5th Edition*. Preparation questions are authored and then reviewed by a group of certified subject-matter experts to ensure the highest level of quality and validity to our product.

If you have any questions concerning this guide or any guide in this series, please visit us on the web at **http://www.trainingbay.cengage.com**.

For online test preparation solutions for ASE certifications, please visit us on the web at **http://www.techniciantestprep.com** to learn more.

ABOUT THE SERIES ADVISOR

Mike Swaim has been an Automotive Technology Instructor at North Idaho College, Coeur d'Alene, Idaho, since 1978. He is an Automotive Service Excellence (ASE) Certified Master Technician since 1974 and holds a Lifetime Certification from Mobile Air Conditioning Society. He served as Series Advisor to all nine of the 2011 Automotive Technician/Light Truck Technician Certification Tests (A Series) of Delmar, Cengage ASE Test Preparation titles, and is the author of *ASE Test Preparation: Automotive Technician Certification Series, Undercar Specialist Designation (X1), 5th Edition.*

The History and Purpose of ASE

ASE began as the National Institute for Automotive Service Excellence (NIASE). It was founded as a non-profit, independent entity in 1972 by a group of industry leaders with the single goal of providing a means for consumers to distinguish between incompetent and competent technicians. It accomplishes this goal through the testing and certification of repair and service professionals. Though it is still known as the National Institute for Automotive Service Excellence, it is now called "ASE" for short.

Today, ASE offers more than 40 certification exams in automotive, medium/heavy duty truck, collision repair and refinish, school bus, transit bus, parts specialist, automobile service consultant, and other industry-related areas. At this time, there are more than 385,000 professionals nationwide with current ASE certifications. These professionals are employed by new car and truck dealerships, independent repair facilities, fleets, service stations, franchised service facilities, and more.

ASE's certification exams are industry-driven and cover practically every on-highway vehicle service segment. The exams are designed to stress the knowledge of job-related skills. Certification consists of passing at least one exam and documenting two years of relevant work experience. To maintain certification, those with ASE credentials must be retested every five years.

While ASE certifications are a targeted means of acknowledging the skills and abilities of an individual technician, ASE also has a program designed to provide recognition for highly qualified repair, support, and parts businesses. The Blue Seal of Excellence Recognition Program allows businesses to showcase their technicians and their commitment to excellence. One of the requirements of becoming Blue Seal recognized is that the facility must have a minimum of 75 percent of their technicians ASE certified. Additional criteria apply, and program details can be found on the ASE website.

ASE recognized that educational programs serving the service and repair industry also needed a way to be recognized as having the faculty, facilities, and equipment to provide a quality education to students wanting to become service professionals. Through the combined efforts of ASE, industry, and education leaders, the non-profit organization entitled the National Automotive Technicians Education Foundation (NATEF) was created in 1983 to evaluate and recognize academic programs. Today more than 2,000 educational programs are NATEF certified.

For additional information about ASE, NATEF, or any of their programs, the following contact information can be used:

National Institute for Automotive Service Excellence (ASE)

101 Blue Seal Drive S.E.

Suite 101

Leesburg, VA 20175

Telephone: 703-669-6600

Fax: 703-669-6123

Website: **www.ase.com**

Overview and Introduction

Participating in the National Institute for Automotive Service Excellence (ASE) voluntary certification program provides you with the opportunity to demonstrate you are a qualified and skilled professional technician who has the "know-how" required to successfully work on today's modern vehicles.

EXAM ADMINISTRATION

> *Note:* After November 2011, ASE will no longer offer paper and pencil certification exams. There will be no Winter testing window in 2012, and ASE will offer and support CBT testing exclusively starting in April 2012.

ASE provides computer-based testing (CBT) exams, which are administered at test centers across the nation. It is recommended that you go to the ASE website at *http://www.ase.com* and review the conditions and requirements for this type of exam. There is also an exam demonstration page that allows you to personally experience how this type of exam operates before you register.

CBT exams are available four times annually, for two-month windows, with a month of no testing in between each testing window:

- January/February – Winter testing window
- April/May – Spring testing window
- July/August – Summer testing window
- October/November – Fall testing window

UNDERSTANDING TEST QUESTION BASICS

ASE exam questions are written by service industry experts. Each question on an exam is created during an ASE-hosted "item-writing" workshop. During these workshops, expert service representatives from manufacturers (domestic and import), aftermarket parts and equipment manufacturers, working technicians, and technical educators gather to share ideas and convert them into actual exam questions. Each exam question written by these experts must then survive review by all members of the group. The questions are designed to address the practical application of repair and diagnosis knowledge and skills practiced by technicians in their day-to-day work.

After the item-writing workshop, all questions are pre-tested and quality-checked on a national sample of technicians. Those questions that meet ASE standards of quality and accuracy are included in the scored sections of the exams; the "rejects" are sent back to the drawing board or discarded altogether.

Depending on the topic of the certification exam, you will be asked between 40 and 80 multiple-choice questions. You can determine the approximate number of questions you can expect to be asked during the Suspension and Steering (A4) certification exam by reviewing the task list in Section 4 of this book. The five-year recertification exam will cover this same content; however, the number of questions for each content area of the recertification exam will be reduced by approximately one-half.

> *Note:* Exams may contain questions that are included for statistical research purposes only. Your answers to these questions will not affect your score, but since you do not know which ones they are, you should answer all questions in the exam.

Using multiple criteria, including cross-sections by age, race, and other background information, ASE is able to guarantee that exam questions do not include bias for or against any particular group. A question that shows bias toward any particular group is discarded.

TEST-TAKING STRATEGIES

Before beginning your exam, quickly look over the exam to determine the total number of questions that you will need to answer. Having this knowledge will help you manage your time throughout the exam to ensure you have enough available to answer all of the questions presented. Read through each question completely before marking your answer. Answer the questions in the order they appear on the exam. Leave the questions blank that you are not sure of and move on to the next question. You can return to those unanswered questions after you have finished the others. These questions may actually be easier to answer at a later time once your mind has had additional time to consider them on a subconscious level. In addition, you might find information in other questions that will help you recall the answers to some of them.

Multiple-choice exams are sometimes challenging because there are often several choices that may seem possible, or partially correct, and therefore it may be difficult to decide on the most appropriate answer choice. The best strategy in this case, is to first determine the correct answer before looking at the answer options. If you see the answer you decided on, you should still be careful to examine the other answer options to make sure that none seem more correct than yours. If you do not know or are not sure of the answer, read each option very carefully and try to eliminate those options that you know are incorrect. That way, you can often arrive at the correct choice through a process of elimination.

If you have gone through the entire exam, and you still do not know the answer to some of the questions, *then guess.* Yes, guess. You then have at least a 25 percent chance of being correct. While your score is based on the number of questions answered correctly, any question left blank, or unanswered, is automatically scored as incorrect.

There is a lot of "folk" wisdom on the subject of test taking that you may hear about as you prepare for your ASE exam. For example, there are those who would advise you to avoid response options that use certain words such as *all, none, always, never, must,* and *only,* to name a few. This, they claim, is because nothing in life is exclusive. They would advise you to choose response options that use words that allow for some exception, such as *sometimes, frequently, rarely, often, usually, seldom,* and *normally.* They would also advise you to avoid the first and last option (A or D) because exam writers, they feel, are more comfortable if they put the correct answer in the middle (B or C) of the choices. Another recommendation often offered is to select the option that is either shorter or longer than the other three choices because it is more likely to be correct. Some would advise you to never change an answer since your first intuition is usually correct. Another area of "folk" wisdom focuses specifically on any repetitive patterns created by your question responses (e.g., A, B, C, A, B, C, A, B, C).

Many individuals may insist that there are actual grains of truth in this folk wisdom, and whereas with some exams this may prove true, it is not relevant in regard to the ASE certification exams.

ASE validates all exam questions and test forms through a national sample of technicians, and only those questions and test forms that meet ASE standards of quality and accuracy are included in the scored sections of the exams. Any biased questions or patterns are discarded altogether, and therefore it is highly unlikely you will experience any of this "folk" wisdom on an actual ASE exam.

PREPARING FOR THE EXAM

Delmar, Cengage Learning wants to make sure we are providing you with the most thorough preparation guide possible. To demonstrate this, we have included hundreds of preparation questions in this guide. These questions are designed to provide as many opportunities as possible to prepare you to successfully pass your ASE exam. The preparation approach we recommend and outline in this book is designed to help you build confidence in demonstrating what task area content you already know well while also outlining what areas you should review in more detail prior to the actual exam.

We recommend that your first step in the preparation process should be to thoroughly review Section 3 of this book. This section contains a description and explanation of the types of questions you will find on an ASE exam.

Once you understand how the questions will be presented, we then recommend that you thoroughly review Section 4 of this book. This section contains information that will help you establish an understanding of what the exam will be evaluating, and specifically, how many questions to expect in each specific task area.

As your third preparatory step, we recommend you complete your first preparation exam, located in Section 5 of this book. Answer one question at a time. After you answer each question, review the answer and question explanation information located in Section 6. This section will provide you with instant response feedback, allowing you to gauge your progress, one question at a time, throughout this first preparation exam. If after reading the question explanation you do not feel you understand the reasoning for the correct answer, go back and review the task list overview (Section 4) for the task that is related to that question. Included with each question explanation is a clear identifier of the task area that is being assessed (e.g., Task A.1). If at that point you still do not feel you have a solid understanding of the material, identify a good source of information on the topic, such as an educational course, textbook, or other related source of topical learning, and do some additional studying.

After you have completed your first preparation exam and have reviewed your answers, you are ready to complete your next preparation exam. A total of six practice exams are available in Section 5 of this book. For your second preparation exam, we recommend that you answer the questions as if you were taking the actual exam. Do not use any reference material or allow any interruptions in order to get a feel for how you will do on the actual exam. Once you have answered all of the questions, grade your results using the Answer Key in Section 6. For every question that you gave an incorrect answer to, study the explanations to the answers and/or the overview of the related task areas. Try to determine the root cause for missing the question. The easiest thing to correct is learning the correct technical content. The hardest things to correct are behaviors that lead you to an incorrect conclusion. If you knew the information but still answered the question incorrectly, there is likely a test-taking behavior that will need to be corrected. An example of this would be reading too quickly and skipping over words that affect your reasoning. If you can identify what you did that caused you to answer the question incorrectly, you can eliminate that cause and improve your score.

Here are some basic guidelines to follow while preparing for the exam:

- Focus your studies on those areas you are weak in.
- Be honest with yourself when determining if you understand something.
- Study often but for short periods of time.
- Remove yourself from all distractions when studying.
- Keep in mind that the goal of studying is not just to pass the exam; the real goal is to learn.
- Prepare physically by getting a good night's rest before the exam, and eat meals that provide energy but do not cause discomfort.
- Arrive early to the exam site to avoid long waits as test candidates check in.
- Use all of the time available for your exams. If you finish early, spend the remaining time reviewing your answers.
- Do not leave any questions unanswered. If absolutely necessary, guess. All unanswered questions are automatically scored as incorrect.

Here are some items you will need to bring with you to the exam site:

- A valid government or school-issued photo ID
- Your test center admissions ticket
- A watch (not all test sites have clocks)

> *Note:* Books, calculators, and other reference materials are not allowed in the exam room. The exceptions to this list are English-Foreign dictionaries or glossaries. All items will be inspected before and after testing.

WHAT TO EXPECT DURING THE EXAM

When taking a CBT exam, as soon as you are seated in the testing center, you will be given a brief tutorial to acquaint you with the computer-delivered test prior to taking your certification exam(s). The CBT exams allow you to select only one answer per question. You can also change your answers as many times as you like. When you select a second answer choice, the CBT will automatically unselect your first answer choice. If you want to skip a question to return to later, you can utilize the "flag" feature, which will allow you to quickly identify and review questions whenever you are ready. Prior to completing your exam, you will also be provided with an opportunity to review your answers and address any unanswered questions.

TESTING TIME

Each individual ASE CBT exam has a fixed time limit. Individual exam time will vary based upon exam area, and will range anywhere from a half hour to two hours. You will also be given an additional 30 minutes beyond what is allotted to complete your exams to ensure you have adequate time to perform all necessary check-in procedures, complete a brief CBT tutorial, and potentially complete a post-test survey.

You can register for and take multiple CBT exams during one testing appointment. The maximum time allotment for a CBT appointment is four and a half hours. If you happen to register for so many exams that you will require more time than this, your exams will be scheduled into multiple appointments. This could mean that you have testing on both the morning and afternoon of the same day, or they could be scheduled on different days, depending on your personal preference and the test center's schedule.

It is important to understand that if you arrive late for your CBT test appointment, you will not be able to make up any missed time. You will only have the scheduled amount of time remaining in your appointment to complete your exam(s).

Also, while most people finish their CBT exams within the time allowed, others might feel rushed or not be able to finish the test, due to the implied stress of a specific, individual time limit allotment. Before you register for the CBT exams, you should review the number of exam questions that will be asked along with the amount of time allotted for that exam to determine whether you feel comfortable with the designated time limitation or not.

As an overall time management recommendation you should monitor your progress and set a time limit you will follow with regard to how much time you will spend on each individual exam question. This should be based on the total number of questions you will be answering.

Also, it is very important to note that if for any reason you wish to leave the testing room during an exam, you must first ask permission. If you happen to finish your exam(s) early and wish to leave the testing site before your designated session appointment is completed, you are permitted to do so only during specified dismissal periods.

UNDERSTANDING HOW YOUR EXAM IS SCORED

You can gain a better perspective about the ASE certification exams if you understand how they are scored. ASE exams are scored by an independent organization having no vested interest in ASE or in the automotive industry. With CBT exams, you will receive your exam scores immediately.

Each question carries the same weight as any other question. For example, if there are 50 questions, each is worth 2 percent of the total score.

Your exam results can tell you:

- Where your knowledge equals or exceeds that needed for competent performance, or
- Where you might need more preparation.

Your ASE exam score report is divided into content "task" areas; it will show the number of questions in each content area and how many of your answers were correct. These numbers provide information about your performance in each area of the exam. However, because there may be a different number of questions in each content area of the exam, a high percentage of correct answers in an area with few questions may not offset a low percentage in an area with many questions.

It should be noted that one does not "fail" an ASE exam. The technician who does not pass is simply told "More Preparation Needed." Though large differences in percentages may indicate problem areas, it is important to consider how many questions were asked in each area. Since each exam evaluates all phases of the work involved in a service specialty, you should be prepared in each area. A low score in one area could keep you from passing an entire exam. If you do not pass the exam, you may take it again at any time it is scheduled to be administered.

There is no such thing as average. You cannot determine your overall exam score by adding the percentages given for each task area and dividing by the number of areas. It does not work that way because there generally are not the same number of questions in each task area. A task area with 20 questions, for example, counts more toward your total score than a task area with 10 questions.

Your exam report should give you a good picture of your results and a better understanding of your strengths and areas needing improvement for each task area.

Types of Questions on an ASE Exam

Understanding not only what content areas will be assessed during your exam, but how you can expect exam questions to be presented will enable you to gain the confidence you need to successfully pass an ASE certification exam. The following examples will help you recognize the types of question styles used in ASE exams and assist you in avoiding common errors when answering them.

Most initial certification tests are made up of between 40 and 80 multiple-choice questions. The five-year recertification exams will cover the same content as the initial exam; however, the actual number of questions for each content area will be reduced by approximately one-half. Refer to Section 4 of this book for specific details regarding the number of questions to expect during the initial Suspension and Steering (A4) certification exam.

Multiple-choice questions are an efficient way to test knowledge. To correctly answer them, you must consider each answer choice as a possibility, and then choose the answer choice that *best* addresses the question. To do this, read each word of the question carefully. Do not assume you know what the question is asking until you have finished reading the entire question.

About 10 percent of the questions on an actual ASE exam will reference an illustration. These drawings contain the information needed to correctly answer the question. The illustration should be studied carefully before attempting to answer the question. When the illustration is showing a system in detail, look over the system and try to figure out how the system works before you look at the question and the possible answers. This approach will ensure that you do not answer the question based upon false assumptions or partial data, but instead have reviewed the entire scenario being presented.

MULTIPLE-CHOICE/DIRECT QUESTIONS

The most common type of question used on an ASE exam is the direct multiple-choice style question. This type of question contains an introductory statement, called a stem, followed by four options: three incorrect answers, called distracters, and one correct answer, the key.

When the questions are written, the point is to make the distracters plausible to draw an inexperienced technician to inadvertently select one of them. This type of question gives a clear indication of the technician's knowledge.

Here is an example of a direct style question:

1. A customer is concerned that it takes increasing effort to steer the car. Which of the following could be the cause?

    A. Worn steering column bushings

    B. Seized steering column U-joint

    C. Worn tie rod end

    D. Loose idler arm

TASK A.1.2

Answer A is incorrect. When steering column bushings wear, they will normally create less friction, not more. The usual customer complaint is a noisy steering column.

Answer B is correct. When the bearings in a U-joint fail, they will often "freeze". This will result in a binding of the steering system as the needle bearings will not travel around the trunnion normally. This problem is often misdiagnosed as a weak power steering pump.

Answer C is incorrect. When a tie rod wears, it usually develops excess movement. This results in a looseness in the steering system, noise, and tie wear.

Answer D is incorrect. A worn idler arm will produce looseness in the steering system causing the vehicle to wander. The customer may comment that they need to continually correct the steering to keep the vehicle on the road.

COMPLETION QUESTIONS

A completion question is similar to the direct question except the statement may be completed by any one of the four options to form a complete sentence. Here's an example of a completion question:

TASK A.2.7

1. The power steering pressure is being measured. The fluid is warm, the engine is idling, the tester valve is open, and the steering wheel is in the straight ahead position. An acceptable reading would be:

 A. 5 psi.
 B. 75 psi.
 C. 200 psi.
 D. 500 psi.

Answer A is incorrect. 5 psi would be too low. The pressure should be between 50 and 150 psi.

Answer B is correct. Pressure under these conditions should normally be more than 50 and less than 150 psi.

Answer C is incorrect. 200 psi would indicate a restriction in the system.

Answer D is incorrect. 500 psi would indicate a restriction in the system.

TECHNICIAN A, TECHNICIAN B QUESTIONS

This type of question is usually associated with an ASE exam. It is, in fact, two true-false statements grouped together, such as: "Technician A says…" and "Technician B says…", followed by "Who is correct?"

In this type of question, you must determine whether either, both, or neither of the statements are correct. To answer this type of question correctly, you must carefully read each technician's statement and judge it on its own merit.

Sometimes this type of question begins with a statement about some analysis or repair procedure. This statement provides the setup or background information required to understand the conditions about which Technician A and Technician B are talking, followed by two statements about the cause of the concern, proper inspection, identification, or repair choices.

Analyzing this type of question is a little easier than the other types because there are only two ideas to consider, although there are still four choices for an answer.

Again, Technician A, Technician B questions are really double true-or-false questions. The best way to analyze this type of question is to consider each technician's statement separately. Ask yourself, "Is A true or false? Is B true or false?" Once you have completed an individual evaluation of each statement, you will have successfully determined the correct answer choice for the question, "Who is correct?"

An important point to remember is that an ASE Technician A, Technician B question will never have Technician A and B directly disagreeing with each other. That is why you must evaluate each statement independently.

An example of a Technician A/Technician B style question looks like this:

1. Technician A say to replace the strut cartridge while replacing the strut bearing. Technician B says to replace the front coil spring while replacing the strut bearing. Who is correct?

    A. A only
    B. B only
    C. Both A and B
    D. Neither A nor B

TASK B.1.12

Answer A is incorrect. The strut cartridge should only be replaced if it is worn.

Answer B is incorrect. The front coil spring should only be replaced if it is worn.

Answer C is incorrect. Both Technicians are incorrect.

Answer D is correct. Neither Technician is correct.

EXCEPT QUESTIONS

Another type of question type used on ASE exams contains answer choices that are all correct except for one. To help easily identify this type of question, whenever it is presented in an exam, the word "EXCEPT" will always be displayed in capital letters. Furthermore, a cautionary statement will alert you to the fact that the next question is different from the ones otherwise found in the exam. With the EXCEPT type of question, only one *incorrect* choice will actually be listed among the options, and that incorrect choice will be the key to the question. That is, the incorrect statement is counted as the correct answer for that question.

Be careful to read these question types slowly and thoroughly; otherwise, you may overlook what the question is actually asking and answer the question by selecting the first correct statement.

An example of this type of question would appear as follows:

1. A vehicle has worn jounce bumpers. All of these could be the cause EXCEPT:

    A. Worn shocks.
    B. Weak springs.
    C. Worn wheel bearings.
    D. Incorrect ride height.

TASK B.1.3

Answer A is incorrect. Worn shocks will cause the suspension to travel further than normal, resulting in worn jounce bumpers.

Answer B is incorrect. Weak springs will allow the vehicle to set lower than normal and result in worn jounce bumpers.

Answer C is correct. Worn wheel bearings can result in noise, but would not cause suspension travel to be abnormal.

Answer D is incorrect. Incorrect ride height could cause the vehicle to set lower than normal, resulting in worn jounce bumpers.

LEAST LIKELY QUESTIONS

LEAST LIKELY questions are similar to EXCEPT questions. Look for the answer choice that would be the LEAST LIKELY cause (most incorrect) of the described situation. To help easily identify these type of questions, whenever they are presented in an exam the words "LEAST LIKELY" will always be displayed in capital letters. In addition, you will be alerted before a LEAST LIKELY question is posed. Read the entire question carefully before choosing your answer.

An example of this type of question is shown here:

TASK A.1.1

1. During inspection of a recirculating ball manual steering gear, which of the following would be LEAST LIKELY to be inspected for wear?

   A.   Sector shaft
   B.   The gear housing
   C.   The rack piston preload
   D.   The pitman arm

Answer A is incorrect. The sector shaft should be checked for wear.

Answer B is incorrect. The gear housing should be checked for wear.

Answer C is correct. The rack piston is found on rack and pinion steering only.

Answer D is incorrect. The pitman arm should be checked for wear.

SUMMARY

The question styles outlined in this section are the only ones you will encounter on any ASE certification exam. ASE does not use any other types of question styles, such as fill-in-the-blank, true/false, word-matching, or essay. ASE also will not require you to draw diagrams or sketches to support any of your answer selections, although any of the described question styles may include illustrations, charts, or schematics to clarify a question. If a formula or chart is required to answer a question, it will be provided for you.

Task List Overview

INTRODUCTION

This section of the book outlines the content areas or *task list* for this specific certification exam, along with a written overview of the content covered in the exam.

The task list describes the actual knowledge and skills necessary for a technician to successfully perform the work associated with each skill area. This task list is the fundamental guideline you should use to understand what areas you can expect to be tested on, as well as how each individual area is weighted to include the approximate number of questions you can expect to be given for that area during the ASE certification exam. It is important to note that the number of exam questions for a particular area is to be used as a guideline only. ASE advises that the questions on the exam may not equal the number listed on the task list. The task lists are specifically designed to tell you what ASE expects you to know how to do and to help prepare you to be tested.

Similar to the role this task list will play in regard to the actual ASE exam, Delmar, Cengage Learning has developed six preparation exams, located in Section 5 of this book, using this task list as a guide. It is important to note that although both ASE and Delmar, Cengage Learning use the same task list as a guideline for creating these test questions, none of the test questions you will see in this book will be found in the actual, live ASE exams. This is true for any test preparatory material you use. Real exam questions are *only* visible during the actual ASE exams.

Task List at a Glance

The Suspension and Steering (A4) task list focuses on six core areas, and you can expect to be asked a total of approximately 40 questions on your certification exam, broken out as outlined here:

A. Steering Systems Diagnosis and Repair (12 questions)
   1. Steering Columns
   2. Steering Units
   3. Steering Linkage
B. Suspension Systems Diagnosis and Repair (12 questions)
   1. Front Suspensions
   2. Rear Suspensions
C. Wheel Alignment Diagnosis, Adjustment, and Repair (11 questions)
D. Wheel and Tire Diagnosis and Service (5 questions)

Based upon this information, the graph shown here is a general guideline demonstrating which areas will have the most focus on the actual certification exam. This data may help you prioritize your time when preparing for the exam.

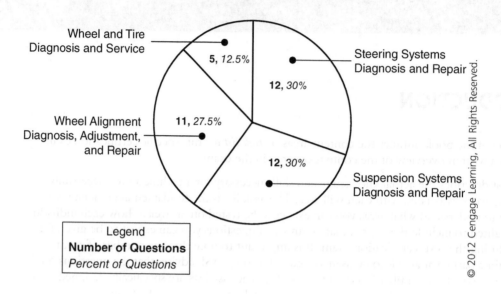

SUSPENSION AND STEERING (A4) TASK LIST

A. Steering Systems Diagnosis and Repair (12 Questions)

1. Steering Columns

1. Diagnose steering column noises and steering effort concerns (including manual and electronic tilt and telescoping mechanisms); determine needed repairs.

The steering column is the direct connection from the steering wheel to the steering gear. Although there are many designs, some basic characteristics apply to most columns. On all new vehicles, there is an airbag located in the center of the steering wheel. Most vehicles will incorporate in the steering wheel and column functions such as turn signals, headlights, hazard lights, ignition lock, horn, ignition switch, wipers, windshield washers, a steering angle sensor, a clock spring assembly, and cruise control switches. Some designs will also have additional features such as radio volume and station switching. A wiring harness runs alongside the column.

To add to driver comfort, many steering columns tilt and may even telescope in and out.

Diagnosing steering complaints always begins with a complete inspection of the steering column. Worn steering column bearings or loose mounts can cause noise and looseness in the steering column. Check for excessive up-and-down and side-to-side movement in the steering wheel and column. A broken support plate or other worn internal column parts can cause looseness or binding in the steering column.

In a tilt steering column, worn or loose pivots, which allow the column to tilt, will result in excessive steering wheel movement. In many cases, it will be necessary to remove the steering wheel and disassemble the steering column to properly identify the worn parts.

2. Inspect and replace steering column, steering shaft U-joint(s), flexible coupling(s), collapsible columns, steering wheels (includes steering wheels and columns equipped with airbags and/or other steering wheel/ column mounted controls, sensors, and components).

Before removing the steering column, point the front wheels in the straight-ahead position and make sure the column is in the lock position. This will prevent damage to the airbag clock spring (spiral cable) and will help line up the steering column with the steering gear during reassembly. For vehicles equipped with airbags, disable the airbag system following the manufacturer's procedure. Usually this involves removing the battery cables or airbag fuse.

After removing the horn pad and/or airbag, remove the steering wheel retaining nut and check alignment marks on the steering wheel and shaft. If no marks are found, scribe marks to ensure proper reinstallation of the steering wheel. Using the proper puller, remove the steering wheel. Never use a hammer, slide-hammer, or knock-off puller to remove the steering wheel. Damage to the column or column bearings can result.

Disconnect any wiring harness connectors, the, bottom steering coupler retaining bolt, and transmission linkage from the column. Mark the position of the steering column coupler to the steering gear. Remove, if necessary, any dash trim to gain access to the steering column mount under the dashboard. Carefully remove the steering column from the vehicle.

Inspect all external components, including the flex coupler and lower column U-joint, if equipped. Many vehicles are designed with an intermediate shaft containing two U-joints. Inspect the U-joints for wear, looseness, and binding. Disassembly of the steering column will be necessary to inspect internal components of the column.

3. Disarm, enable, and properly handle airbag system components during vehicle service following manufacturers' procedures.

The supplemental inflatable restraint system (airbag) must be disarmed properly before any work is performed on the steering column. Failure to disarm the airbag may result in the accidental deployment of the airbag, which can cause personal injury and unnecessary repairs to the airbag system. Always consult the service manual for recommended procedures on disabling the airbag system.

The following is a list of basic procedures and precautions that should be followed when you work with and around airbag systems:

1. Always disable the airbag system before servicing the airbag or any components near or around the steering column and/or airbag system.
2. Never subject the inflator module to temperatures greater than 175°F (79°C).
3. If any airbag component is dropped, it should be replaced.
4. Never test any airbag component with electrical test equipment unless instructed to do so by the factory service manual.

5. When carrying a live (not deployed) airbag, always point the bag and trim cover away from you.

6. When placing a live airbag on a workbench, always face the airbag and trim cover up.

7. Discarded live airbags must be deployed. Follow manufacturer's procedure for proper disposal.

8. Lock the steering wheel in place whenever removing the steering wheel, steering column, or steering gear, to prevent damage to the clock spring. The clock spring maintains a continuous electrical connection as the steering wheel rotates between the inflator module and airbag controller.

4. Diagnose, inspect, adjust, repair or replace components (including motors, sensors, switches, actuators, harnesses, and control units) of steering column-mounted, electronically controlled, hydraulically and/or electrically assisted steering systems; initialize systems as required.

All column mounted electric steering systems use an electric motor, which is attached to and operates the steering linkage between the steering wheel and the rack and pinion unit. The ECU and electric motor will typically be mounted under the dash. A variety of motor designs and gear drives are possible, depending on the vehicle. Steering effort can be controlled by reducing steering assist while driving straight ahead at highway speeds and increasing steering assist at low speeds.

Typical electronic steering systems incorporate a microprocessor, steering sensor, electric motor, vehicle speed sensor, differential sensor, and other sensors. Inputs from the sensors are sent to the steering control unit and evaluated. Depending on the input from various sensors, the computer will send out the appropriate command to the power unit, which will then control the steering motor. Steering assist will change as the vehicle increases and decreases speed and when the vehicle is turning left or right.

Diagnosing electronically controlled steering starts with a complete visual inspection of the unit, motor, and sensors. Electronic steering systems have the capability of storing diagnostic trouble codes, which will help in analyzing a fault within the system. If a problem does occur, the control unit will shut down the system and employ a fail-safe mode. Steering will revert to manual mode. A warning light will illuminate on the dash panel to alert the driver. Follow the appropriate service manual for diagnostic procedures and for obtaining trouble codes.

2. Steering Units

1. Diagnose steering gear (non-rack and pinion type) noises, binding, vibration, freeplay, steering effort, steering pull (lead), and leakage concerns; determine needed repairs.

Conventional steering gears utilize the recirculating ball design. Many conventional steering gear problems are due to internal wear. Start your steering inspection by turning the steering wheel full left and full right, noting any noise, binding, roughness, looseness, or excessive effort. Check lubrication level and inspect the steering gear for indications of fluid leakage from the gear, and at hose connections for power steering units. Loss of fluid may cause increased steering effort and erratic steering. Fluid loss on power steering units may cause a whine or growling noise. A loose or worn power steering belt may cause a squeal while turning the steering wheel, and it may also cause erratic steering and increased steering effort. On hydraulically assisted power steering systems, a missing belt will result in loss of power steering assist.

A worn sector shaft, wormshaft, or bearings can cause binding and roughness while the steering wheel is turned. Excessively worn steering gears will have to be either overhauled or replaced.

Inspect the steering gear mounting bolts. Loose gear-to-frame mounting bolts will cause excessive freeplay and wandering and may even cause a vibration. Inspect the frame where the steering gear is mounted for corrosion and cracks. Rust can weaken the frame and cause the gear to flex or pull away, causing a potentially dangerous situation.

2. Diagnose rack and pinion steering gear noises, binding, vibration, freeplay, steering effort, steering pull (lead), and leakage concerns; determine needed repairs.

Most new vehicles are equipped with rack and pinion steering gear. The rack and pinion steering gear is a much simpler design. The steering column is connected directly to the sector shaft, called the pinion. The pinion operates the rack assembly, which moves the tie rods and the steering knuckles. On hydraulically assisted rack and pinion gears, a control valve directs the flow of fluid from one side of the rack to the other. On electric-assisted gears, an electric motor assists the gear.

Diagnose problems with rack and pinion steering gear following the same basic procedures as with the conventional steering gear. Turn the steering wheel full left and full right, noting any noise, binding, roughness, looseness, freeplay, or excessive effort. Inspect the mounting brackets and bushings for wear, which can cause loose or wandering steering. A binding or roughness while turning the wheel may indicate a worn rack gear. A worn steering rack may also cause excessive steering effort. Carefully inspect the rack for leaks at the bellows/boots and at the pinion seal. Electric-assist rack and pinion steering gears may need to be recentered after a four-wheel alignment. This is done using a scan tool. Failure to center the rack in this manner may result in a pull and/or a crooked steering wheel.

3. Inspect power steering fluid level and condition; determine fluid type and adjust fluid level in accordance with vehicle manufacturers' recommendations.

Power steering fluid reservoirs are either integrated with the pump or remotely located. Many remote reservoirs are transparent and have markings on the outside to indicate the fluid level. Remote reservoirs are usually marked FULL COLD, FULL HOT, or MIN, MAX. If the fluid is checked with a dipstick, wipe the cap and area clean to prevent dirt from entering the system.

Check the fluid level at normal operating temperature (approximately 175°F or 79°C). Make sure the fluid level is at least MIN or COLD before running the engine to prevent damage to the pump, or causing air to be whipped into the fluid creating noise. A low fluid level may indicate a leak.

When checking the fluid level, note the fluid condition. Discolored fluid or signs of particles may be an indication of wear to the pump, hoses, or gear. Foamy fluid is caused by air in the system.

Before adding fluid, check with the factory manual for the proper fluid.

4. Inspect, adjust, align, and replace power steering pump belt(s), tensioners, and pulleys.

The power steering pump is driven by either a V-belt or serpentine belt. Most new vehicles are equipped with serpentine belts that incorporate automatic tensioners. Older belt system designs must be adjusted manually. Belt alignment is critical for silent operation. Always check for proper pulley alignment when inspecting or replacing belts. Belt alignment can be checked with a laser belt alignment tool. If the belt is not in correct alignment, usually the pulley on the pump can be moved in or out to align correctly.

Belts should be inspected for glazing, rotting, cracking, or swelling. Oil residue on the belt may be an indication of a fluid leak. Belt tension must also be checked on both automatic tensioners and manually adjusted belts. Use a belt deflection gauge or use the deflection method by applying pressure with your finger on the belt midway between the longest span. Compare your reading with manufacturer's specifications. Typical belt deflection is approximately 0.5 inch (12.7mm).

For automatic tensioners, inspect for wear, looseness, and binding. Also inspect the tensioner pulley bearing for wear, looseness, and binding. If equipped, also inspect the idler pulley bearing for wear looseness, and binding. Replace if worn.

5. Diagnose power steering pump noises, vibration, and fluid leakage; determine needed repairs.

The power steering pump supplies the hydraulic pressure needed to operate the steering gear. The most common problem with the pump is seal leaks. Loss of fluid will cause an audible whine or growling noise. Seal leaks will require either pump replacement or an overhaul. If a vibration is felt when turning the steering wheel, check for belt problems, internal pump problems, or loose pump retaining bolts and mounting brackets.

6. Remove and replace power steering pump; inspect pump mounting and attaching brackets; remove and replace power steering pump pulley; transfer related components.

To remove the power steering pump, start by removing the belt and cleaning any dirt around the pump and hose connections. Disconnect the hoses and plug the hoses and pump fittings. Disconnect any electrical connectors from variable assist solenoid, or pressure sensor (if equipped). Remove the pump mounting bolts and carefully remove the pump from the vehicle. On some vehicles, it will be necessary to remove the pulley before the pump can be removed from the mounting bracket. To remove and install the pulley, always use the proper puller. Never attempt to hammer the pulley off or on. With the pump removed, inspect the mounting brackets.

After installing the pump, fill with fluid and bleed any air from the system per the vehicle manufacturer's procedure, and then check for leaks. Also check belt tension and for proper pulley alignment. Road test the vehicle and recheck fluid level and for leaks.

7. Perform power steering system pressure and flow tests; determine needed repairs.

Increased power steering effort or lack of power steering assist can be caused by problems in the pump or gear. Checking system pressure can help determine the cause of the problem.

Before starting pressure testing, check fluid level and fill to proper level. Run the engine until normal power steering fluid operating temperature is achieved. Make sure the engine idle is correct and check belt tension. Obtain manufacturer's specifications for the vehicle being tested.

Install a power steering pressure gauge tool on the pressure side between the pump and the gear. Position the shutoff valve toward the power steering gear. Start the engine and record the pressure in the straight-ahead position with the gauge valve open. If the pressure reading is above specification, check for restricted hoses or a damaged steering gear.

Next, turn the steering wheel full left or full right and hold the wheel against the stop for no more than five seconds. Record the maximum pressure attained and check factory specification. Typical pressure readings are generally over 1,000 psi, 6894.745 kPA. If the pressure reading is below specification, position the steering wheel in the straight-ahead position and slowly close the shutoff valve. If the pressure rises to the proper specification with the shutoff valve closed, the pump is working correctly. Therefore, the reason for low pressure while turning the steering wheel lies in the gear assembly. If the pressure does not rise with the shutoff valve closed, the problem is in the pump. Do not leave the shutoff valve closed for more than five seconds.

8. Inspect and replace power steering hoses, fittings, o-rings, coolers, and filters.

Pressure side power steering hoses are designed to withstand the extreme high pressure developed by the power steering pump and the high temperatures generated under the hood. Inspect power steering hoses for leaks, cracks, swelling, and physical damage. Always replace a power steering hose with one specifically designed for the vehicle being worked on. Hoses must be routed and mounted correctly.

Some hose fittings use o-ring seals. Always replace the o-ring when replacing a power steering hose. Tighten the new hose to correct specification and check for leaks.

Many power steering systems use coolers to maintain proper fluid temperature. Inspect coolers for leaks and damage. A restricted cooler can cause fluid overheating and pump and gear damage. Filters are often used to trap particles that may damage internal components.

9. Remove and replace steering gear (non-rack and pinion type).

The following are the basic steps to remove and replace a power steering gear:

1. Determine if the steering column must be removed or loosened inside the vehicle. If it must be removed or loosened, proceed as follows. If not, proceed with undercar and underhood operations beginning with step 2.

   a. Disconnect the battery ground cable or remove the airbag fuse(s), if equipped. If the vehicle has airbags, wait two to five minutes before proceeding.

    b. Disconnect electrical connectors from the steering column under the instrument panel.

    c. Loosen or remove the steering column mounting bolts from the instrument panel bracket.

2. Disconnect the power steering hoses from the steering gear box and drain the fluid into a suitable container; dispose of properly.

3. If necessary for access, remove the power steering pump and any other engine-driven accessories, as required.

4. Make alignment marks for reassembly and disconnect the steering column from the steering gear box. The steering wheel should be locked in place to prevent rotation. If the wheel is rotated, the clock spring could be damaged.

5. Remove the pitman arm from the gear box sector shaft, using a suitable puller.

6. Unbolt the steering gear box from the chassis and remove it from the vehicle.

7. Reinstallation is the reverse of removal. Properly torque all retainers to specification.

8. After installing the steering gear box and reconnecting the hoses, fill the pump with fluid and bleed air from the system.

10. Remove and replace rack and pinion steering gear; inspect and replace mounting bushings and brackets.

1. Before removing the rack and pinion gear from the vehicle, point the front wheels in the straight-ahead position.

2. If the vehicle is equipped with an airbag, it is important to lock the steering wheel in place. If the steering wheel rotates while not connected to the gear, damage will occur to the airbag clock spring.

3. Disconnect the steering column coupling or U-joint from the steering gear. Mark the position of the steering column shaft and steering gear to ensure correct alignment during reassembly.

4. Disconnect the power steering fluid lines from the steering gear (if equipped) and drain the fluid into a suitable container.

5. Remove the outer tie rods from the steering knuckle and remove the mounting bolts. The rack and pinion will be bolted either to the frame or to the engine cradle.

6. Remove the rack and pinion gear from the vehicle and carefully inspect for signs of damage and leakage.

7. Check for wear in the rack pinion and inner tie rods. Inspect the rack bellows/boots.

8. Inspect the mounting brackets and bushings for wear. Replace worn or damaged bushings and brackets.

9. When reinstalling the rack, torque all retainers to specification. If the engine cradle was loosened, make sure it is aligned properly during reassembly.

11. Adjust steering gear (non-rack and pinion type) worm bearing preload and sector lash.

Worm bearing preload affects steering effort. Too little bearing preload and the steering will feel loose. Too much preload and the steering will feel tight or stiff. To check worm bearing preload, remove the pitman arm and the horn pad assembly. With an inch pound torque wrench, slowly rotate the steering wheel to the left and to the right by using the steering wheel retaining nut. Some manufacturers may require the measurement to be taken with the torque wrench at the steering gear input shaft. Note the reading while rotating the steering wheel and compare with manufacturer's specification. If an adjustment is needed, loosen the worm shaft bearing locknut. To increase bearing preload, tighten the adjuster nut. To decrease preload, loosen the adjuster nut. After the adjustment is made, turn the steering wheel to the left and to the right. If any roughness or binding is felt, the steering gear may be worn and may have to be replaced. Some systems use shims to adjust preload. Removing shims will increase preload; adding shims will decrease preload.

Sector shaft lash adjusts the contact between the recirculating ball nut and sector shaft within the gearbox. To check sector lash, remove the pitman arm and horn pad as before, and determine the exact center of the steering gear. Rotate the steering wheel to the left one-half turn. Using an inch pound torque wrench, measure the steering wheel's resistance as it is turned to the right, passing the center point, and continuing until the steering wheel is rotated one full turn. Check your reading against factory specification. If adjustment is required, loosen the sector shaft adjusting locknut and turn the adjusting screw as required. Note: Some manufacturers may require the gear to be drained of oil prior to these adjustments. Other manufacturers may require the steering linkage be disconnected from the steering gear when this measurement is performed.

12. Inspect and replace steering gear (non-rack and pinion type) seals and gaskets.

Leaking steering gears often mean that there is considerable wear, and replacement of the gear is required. Carefully inspect the steering for excessive wear before recommending seals or gasket replacement. Consult the factory manual for procedures and specifications. The steering gear may have to be removed from the vehicle in order to replace some seals.

Thoroughly clean the exterior of the gear before starting the repair. Carefully disassemble the section of the steering gear being repaired and inspect and clean all parts. Replace worn seals and gaskets as needed. Use appropriate tools to install seals as required. After reassembly, check and perform any necessary adjustments and check for leaks.

13. Adjust rack and pinion steering gear.

Two adjustments may be possible on a rack and pinion assembly:

1. Pinion torque is the force needed to turn the pinion gear along the rack. It is adjusted by turning an adjustment screw or a threaded cover on the rack housing, or by adding or removing shims under the rack support cover. This adjustment is also known as the rack yoke (bearing) adjustment.

2. Pinion bearing preload is the force that the pinion bearings place on the pinion shaft. Only a few steering assemblies have adjustable pinion bearing preload. When it is adjustable, adjustments are made by adding or removing shims or by turning an adjustment collar at the base of the pinion gear.

Most vehicles will require that the rack and pinion assembly be removed for adjustment. On some vehicles, you may be able to disconnect the steering shaft and the tie rods and make the adjustments on the car. The steering shaft and tie rods must be disconnected to remove all steering load from the rack and pinion assembly.

14. Inspect and replace rack and pinion steering gear bellows/boots.

The rack and pinion steering gear has a bellows-type boot at each end of the rack assembly to protect the inner tie rod ball sockets and rack seals from dirt and moisture. The boots contract and expand with the turning of the wheels. Inspect the boots for wear, tears, and for fluid seepage. Leaking fluid is an indication of a defective seal or worn rack and pinion. If a boot is cracked, dirt and moisture may have entered. Inspect the inner tie rod for wear.

To replace the boots, it will be necessary to remove the outer tie rod and tie rod locking nut on some models. Mark the position of the tie rod in order to maintain proper toe alignment during reinstallation. Remove the bellows/boots retaining clamps and slide the boot off. Replace the boots using new retaining clamps. When the tie rod is removed, it is important that the wheel alignment be checked after reassembly.

15. Flush, fill, and bleed power steering system.

Flushing the power steering system is accomplished by disconnecting the return to the pump. Plug the pump return port and fill the power steering reservoir with the recommended fluid. With the front wheels off of the ground and the return hose in a drain pan, start the engine and slowly turn the steering wheel from stop to stop. Flushing the system with two quarts of power steering fluid should be sufficient to remove all contaminants and foreign material. Cleaning solvent should never be used in power steering systems for cleansing or flushing procedures.

After a repair has been made to the power steering system, bleeding will be necessary to remove trapped air and to obtain a correct fluid level. Refer to the factory manual for bleeding procedures on the specific vehicle being serviced.

16. Diagnose, inspect, repair, or replace components of variable-assist and/or variable ratio steering systems steering systems.

In a variable-assist and/or variable ratio steering system with a steering wheel rotation sensor, the hydraulic boost increases when the steering wheel rotation exceeds a specified limit. Power steering assist also is increased at low speeds and decreased at higher speeds.

The system is designed to provide better feel and control at higher vehicle speeds. The variable steering systems are usually designed to start firming up the steering at speeds over 25 mph (40 km/h) and to reach the maximum firmness between 60 and 80 mph (97 and 129 km/h), depending on design. On most vehicles, the main input for the variable-assist steering systems is the vehicle speed sensor, but some manufacturers also use a steering wheel rotation sensor so the vehicle will revert to full assist during evasive maneuvers. On most vehicles, the system goes to full assist below 25 mph (40 km/h).

Pump pressure is controlled by a variable orifice, or pressure control valve. Hydraulic pressure is reduced, or gradually reduced, as vehicle speed increases. Problems within the system may cause a noticeable change in the amount of steering effort at different speeds. The steering may feel heavier at low speeds or lighter at high speeds. A faulty speed sensor will prevent the system from operating properly. On computerized systems, the controller will put the system in full-assist mode at all speeds and steering maneuvers if a fault is detected. If the system is computer controlled, scan the control module for any stored trouble codes as part of your diagnosis.

17. Diagnose, inspect, adjust, repair or replace components (including motors, sensors, switches, actuators, harnesses and control units) of rack mounted, electronically controlled, hydraulically and/or electrically assisted steering systems; initialize systems as required.

Rack mounted electric power steering units differ from steering column mounted electric power steering units in that assist is applied to the rack, not the steering column linkage. All rack mounted electric power steering systems use an electric motor, which is attached to and operates the steering rack. A variety of motor designs and gear drives are possible, depending on the vehicle. The electric motor may be connected to the rack through a gear box, or be an integral part of the rack with the motor wrapped around the rack unit. The electric motor may operate on 12 or 42 volts. As with the electro-hydraulic steering system, steering effort can be controlled by reducing steering assist while driving straight ahead at high speeds and increasing steering assist at low speeds.

Typical electronic steering systems incorporate a microprocessor, steering sensor, electric motor, vehicle speed sensor, differential sensor, and other sensors. Inputs from the sensors are sent to the steering control unit and evaluated. Depending on the input from the various sensors, the computer will send out the appropriate command to the power unit, which will then control the steering motor. Steering assist will change as the vehicle increases and decreases speed and when the vehicle is turning left or right.

Diagnosing electronically controlled steering starts with a complete visual inspection of the unit, motor, and sensors. Electronic steering systems have the capability of storing diagnostic trouble codes, which will help in analyzing a fault within the system. If a problem does occur, the control unit will shut down the system and employ a fail-safe mode. Steering will revert to manual mode. A warning light will illuminate on the dash panel to alert the driver. Follow the appropriate service manual for diagnostic procedures and for obtaining trouble codes.

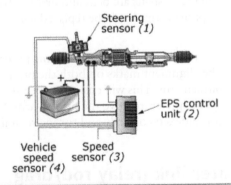

1. Sensor de la dirección
2. Módulo de control de la dirección eléctrica asistida (EPS)
3. Sensor de velocidad
4. Sensor de velocidad del vehículo (VSS)

Electric power steering system

3. Steering Linkage

1. Inspect and adjust (where applicable) front and rear steering linkage geometry (including parallelism and vehicle ride height).

The two most common steering linkage designs in use today are the rack and pinion system and the parallelogram design. The parallelogram design is a much more complicated system that uses a conventional steering gear box.

A variation of the parallelogram design is the cross steer linkage system, used on some four-wheel drive vehicles. The cross steer linkage uses one long tie rod, which is connected to both the left and right steering arms. A drag link connects the left steering arm to the steering gear box. Some cross steer systems use an adjustable tie rod drag link, which is used to center the steering wheel.

On all vehicles, a complete periodic inspection of the steering linkage is required due to normal wear and tear of the steering components. All nuts and cotter pins should be in place, and you should inspect for leaking grease seals. Excessive wear in a steering linkage component will result in inadequate steering performance, the feeling of looseness in the steering wheel, and can cause tires to wear unevenly.

The steering linkage should also be checked for bent or damaged components. A bent component, such as a tie rod or center link, can affect steering ability and may cause irregular tire wear. Vehicle ride height must also be checked when performing a routine linkage inspection. On some vehicles equipped with torsion bars, ride height can be adjusted. On most vehicles, if ride height is incorrect, suspect worn or broken springs. Some vehicles are equipped with rear linkage systems and are susceptible to the same problems as in the front linkage system. A complete inspection of the rear linkage system should be performed as with the front.

2. Inspect and replace pitman arm.

The main reason for changing pitman arms is vehicle crashes. Pitman arms are very well built, and the splines do not loosen from road shock. The pitman arms can be bent or broken in vehicle crashes, but more often sector shafts are bent and broken. If the pitman arm has a ball and socket at one end, the pitman arm should be replaced when the ball and socket show any looseness.

To remove the pitman arm, remove the nut and lock washer first. Some pitman arms are indexed to the sector shaft. If not, scribe alignment marks on both the pitman arm and the sector shaft before removing the pitman arm. This will ensure proper alignment during reinstallation. Using the appropriate puller, remove the pitman arm. When installing the pitman arm, make sure it is indexed correctly and torque retaining nuts to specification.

3. Inspect and replace center link (relay rod/drag link/intermediate rod).

A bent or damaged center link can cause incorrect toe alignment and front-wheel shimmy and may cause tire wear. Inspect the center link for worn joints and for physical damage. To remove the center link, remove the cotter pins and nuts and separate the joints using a separator fork or puller. After installing the center link, properly torque all retaining nuts and replace the cotter pins. A wheel alignment should be performed after replacement.

4. Inspect, adjust (where applicable), and replace idler arm(s) and mountings.

The function of the idler arm is to hold the right side of the center link level with the left side of the steering linkage. The idler arm is set at the exact same angle and is the same length as the pitman arm. The weakest point on the idler arm is the bushing. A worn idler arm bushing will cause excessive movement in the arm, which can cause changes in toe alignment and loose steering. Some idler arms have replaceable bushings; most will have to be replaced if wear is excessive. Wear should be checked with hand pressure only; never use large pliers or extreme pressure.

To remove an idler arm, disconnect the arm from the steering linkage and unbolt from the frame. Some idler arms have adjustable slots in the frame. Mark the idler arm prior to removing to ensure proper positioning when reinstalling. If the idler arm is installed in the incorrect position, the center link will be unlevel and bump steer can result.

5. Inspect, replace, and adjust tie rods, tie rod sleeves/adjusters, clamps, and tie rod ends (sockets/bushings).

The tie rods connect the center link to the steering knuckle on conventional steering systems. On rack and pinion steering, the tie rods connect the steering knuckle to the steering gear. Conventional steering systems have inner and outer tie rods coupled by a sleeve on both sides. The inner tie rods connect to the center link. On rack and pinion designs, the inner tie rods connect to the steering gear. Toe alignment adjusters are located on the tie rod assembly. Toe must be checked after replacement of a tie rod.

Worn tie rods can cause front end shimmy, loose steering and incorrect toe, and may cause tires to wear. The best method to find worn tie rod ends is during a dry park inspection. With the wheels on the ground, have an assistant turn the steering wheel back and forth slowly while you check for loose parts. Inspect the tie rods for excessive up and down movement at the ball joint socket. Also check for movement where the toe rod threads into the adjustment sleeve. Tie rods with excessive movement will cause changes in toe and affect steering performance and should be replaced.

The tie rod sleeves must be rotated to adjust front-wheel toe and center the steering wheel.

Replacing the inner tie rod ends (a common repair) must be done carefully to prevent damage to the pinion teeth on rack and pinion steering. The rack must be held firmly while the socket threads are loosened from the threads of the rack.

If the tie rod sleeve clamp is not positioned correctly before being tightened, the clamp will not exert enough force to hold the threads together. The constant motion while the vehicle is in operation will wear the threads involved, and the two pieces will pull apart, resulting in loss of steering control. On some vehicles, if the tie rod sleeve clamp is not positioned with the proper orientation, the sleeve bolt could rub against a crossmember or a suspension part, or wear through a power steering hose.

6. Inspect and replace steering linkage damper(s).

Steering linkage dampers (sometimes called steering linkage shock absorbers) are found on vehicles with parallelogram steering systems. Vehicles with solid front axles, large wheel/tire combinations and/or high caster settings will most often need a steering damper. The damper is secured on one end to the vehicle frame and is connected at the other end to the center link.

Steering linkage dampers are designed to help the steering and suspension systems keep road shock under control. High-speed vibrations are caused mostly by tire and wheel imbalance. Steering linkage dampers work like shock absorbers and are checked in the same way one would check for a bad shock.

B. Suspension Systems Diagnosis and Repair (12 Questions)

1. Front Suspensions

1. Diagnose front suspension system noises, handling, ride height, and ride quality concerns; determine needed repairs.

The most common front suspension complaints are noise, steering pull, irregular tire wear, excessive body roll, poor ride quality, and wheel shimmy. These problems are usually the result of worn bushings, worn ball joints, weak springs, faulty shock absorbers, defective tires, or worn steering linkage components. Broken springs or shocks will affect body roll and cause noise, and may affect steering ability. Leaking shocks will not affect vehicle ride height but may cause poor ride quality.

Road testing the vehicle may reveal abnormal noises, pulling, or steering problems. Pushing down at each corner of the vehicle may uncover noises caused by worn shock bushings, broken springs, or worn suspension components. Wheel bearings and tires should not be overlooked as sources of problems. An out of round tire or loose wheel bearing may affect steering handling and ride quality. Also, do not overlook a faulty tire when diagnosing a pull to one side.

Vehicle ride height should be checked on all vehicles and compared to factory specifications. Incorrect ride height is usually caused by broken or worn springs. Some vehicles are equipped with adjustable torsion bars to correct ride height. Before adjusting the torsion bars, carefully inspect the suspension for worn or bent components.

Other common sources for noise are worn ball joints and control arm bushings. As the vehicle travels over bumps, a squeak or groan can be heard. If the vehicle is equipped with a sway bar, inspect the sway bar links, link bushings, and frame mounts. Broken sway bar links will cause a cracking or banging noise and may cause excessive body roll.

2. Inspect and replace upper and lower control arms, bushings, and shafts.

Control arms allow for the upward and downward movement of the suspension and wheels. The control arms pivot at the ball joints and at the frame. Bushings at the frame allow for this movement. Inspect both lower and upper control arm bushings for wear. Removal of the control arms will be necessary to replace bushings. Some upper control arm designs incorporate a shaft and bushing assembly. This shaft must be carefully inspected and replaced if worn. Control arm bushings are usually replaced with the use of a press. After new bushings are installed, the fasteners are not tightened until the vehicle is on the ground; normal weight of the vehicle is on the suspension and sitting at its normal ride height.

Place safety stands under the lower control arms near the ball joints on most vehicles when replacing the upper control arms because the springs must be partially compressed. When you replace the lower control arms, the safety stands must support the vehicle by the frame so the arms can move down while the springs are removed. The safety stands must be placed in different positions under the vehicle when you replace upper control arms than when replacing lower control arms because of spring location.

3. Inspect and replace rebound and jounce bumpers.

When the wheel hits a bump, the control arms pivot upward, causing the spring and shock to compress. Rubber bumpers cushion the blow if the control arms reach their limit of travel. Inspect the rebound or jounce bumpers for wear and cracks. In some cases, the bumper may actually be missing. Jounce bumpers that are worn, cracked, or missing may be caused by worn springs which lowers the vehicle ride height, which in turn causes the control arms to reach the limit of travel. Inspect for this condition and replace the springs if needed.

4. Inspect, adjust, and replace track bar, strut rods/radius arms, and related mounts/bushings.

Since a strut rod/radius arm (compression/tension) positions the lower control arm, worn strut rod bushings or a bent strut rod can cause changes in caster, camber, and toe. Worn bushings may result in the lower control arm moving rearward during braking.

Worn strut rod bushings can cause the vehicle to pull to the direction of the worn bushing every time the brakes are applied. Worn strut rod bushings can cause alignment problems.

5. Inspect and replace upper and lower ball joints (with or without wear indicators).

When a coil spring is mounted between the lower control arm and the chassis, a jack must be positioned under the lower control arm to unload the ball joints.

Do not place the safety stands under the frame to check for play in the load-carrying ball joint, because the front spring tension would make it impossible to measure actual freeplay. Placing the safety stands under the lower arms on vehicles equipped with MacPherson struts would make ball joint freeplay impossible to measure because the ball joints would be supporting the weight of the front of the vehicle.

Ball joints carrying the majority of the vehicle weight are known as load-carrying ball joints. After correctly positioning the jack stand, check the ball joints for excessive wear. Consult the factory manual for specifications. Some ball joints are designed with a wear indicator at the threaded portion of the grease fitting. If the grease fitting shoulder is receded flush with the outer surface of the ball joint, the ball joint must be replaced.

Worn ball joints can cause alignment problems, tire wear, hard steering, and wheel shimmy. On vehicles with upper and lower ball joints, the load-carrying ball joint usually wears first, but always inspect both. Ball joints must be properly unloaded to check for wear. When the coil spring is located between the lower control arm and the frame, position a jack under the lower control arm. Raise the jack until there is clearance between the floor and the tire. The ball joint is now unloaded. Wear is determined by checking the amount of movement of the ball joints. Check the appropriate service manual for vehicle specifications. Many alignment experts recommend no perceptible movement as a specification. On MacPherson strut designs, raise the vehicle until the tire is off the ground by the frame and let the lower control arm hang down.

To replace a ball joint, make sure the control arm is properly supported on either the control arm or frame, so that the ball joint is not under tension from the spring.

Remove the cotter pin (if equipped) and the ball joint retainer nut. Using a ball joint fork, separate the ball joint. Ball joints can be threaded, riveted, pressed, or bolted in place on the control arm. Use the correct tool, depending on the design. On some newer vehicles, the ball joints cannot be serviced separately and may require replacement of the control arm.

A new threaded ball joint must be torqued to specification. A riveted ball joint will be replaced with a new ball joint supplied with a hardware kit containing bolts, nuts, and washers. Torque the nuts or bolts accordingly. Tighten the ball joint retaining nut to specification and replace the cotter pin, if so equipped. If the hole does not align to install the cotter pin, continue tightening the nut until it does; do not loosen the nut to align the hole. Grease the ball joint if it is equipped with a grease fitting.

6. Inspect non-independent front axle assembly for damage and misalignment.

Non-independent front axles are primarily used on some four-wheel drive trucks and heavy-duty one ton two-wheel drive trucks. It is a simple and strong design, requiring little maintenance. The non-independent front axle incorporates a solid front axle and is fitted with either leaf or coil springs. The front wheels steer by pivoting on kingpins or ball joints, located on the spindle at the ends of the solid axle. Realignment is necessary only if parts are bent or damaged.

A disadvantage of the non-independent front axle is that it provides a rougher ride than independent front suspensions. The up-and-down movement of the front wheels tends to cause a tipping effect and imposes a twisting motion to the frame.

Although they are durable, a periodic inspection is required on non-independent front suspensions. Inspect the axle for warpage, bending, twisting, cracks, and physical damage. Damage may not be obvious, and looking for abnormal tire wear may sometimes identify a problem with the suspension. Misalignment of the axle will cause tracking problems and tire wear.

Inspect the king pins or ball joints for wear. If the assembly is equipped with leaf springs, check for proper alignment of the springs to the front axle. If the vehicle design uses coil springs, check for worn or broken springs and the radius arms and bushings for wear. If the radius arm bushings are badly worn, carefully inspect the radius arm bracket for signs of excessive wear at the point where the radius arm passes through the bracket.

7. Inspect and replace front steering knuckle/spindle assemblies and steering arms.

Excessive tire squeal while cornering may be caused by improper turning angle, also known as toe-out-on-turns . This problem can be caused by a bent steering arm. Bent steering arms and steering knuckles or spindles show up in the alignment readings for toe-out-on-turns and in the steering axis inclination readings. Sometimes a technician can see rust flakes or disturbed metal at the bent section of the part. The parts named must be replaced if they are bent or otherwise damaged.

To remove the steering knuckle, raise the vehicle and remove the wheel and brake components. Support the suspension so that all the tension is removed from the ball joint (s). Disconnect the tie rod end and separate the ball joints. On some MacPherson strut designs, the strut to knuckle bolts are used to adjust camber. Mark the position of the bolts so proper camber can be maintained during reinstallation.

To install the steering knuckle, reverse the procedure, torque all fasteners to specification, and check the front-wheel alignment. Road test the vehicle after installation to ensure proper brake performance and steering operation.

8. Inspect and replace front suspension system coil springs and spring insulators (silencers).

Coil springs are located between the axle, or control arm, and the frame or incorporated on a strut and shock assembly. The spring will compress to support the vehicle at a specific ride height. The spring will also compress and rebound in a controlled manner as the vehicle travels over uneven roads. As the springs wear, ride height and ride quality will diminish. Inspect the vehicle for correct ride height. Incorrect ride height would indicate a weak or broken spring. Also, check the insulators at the top or bottom of the spring seats. On a MacPherson strut design, check the upper strut mount and bearing. Always replace springs in pairs, either both front, or both rear.

To remove a coil spring, raise the vehicle off the ground and remove the wheel. Disconnect all steering and brake components necessary to gain access to the spring. Compress the spring using the appropriate spring compressor. Support the lower control arm and separate the lower ball joint. It may be necessary to remove the shock absorber. Lower the control arm and remove the spring. With the spring removed, inspect insulators and spring seats for wear. The new spring will have to be compressed to be installed. Make sure the new spring is positioned correctly, so that it sits correctly on the control arm and/or spring seat. Reinstall all components, check the front-end alignment, and road test the vehicle. On MacPherson strut designs, the entire strut and shock assembly is removed as a unit from the vehicle. The spring is compressed, and the upper spring mount is removed. Remove the spring and inspect the upper mount, strut bearing, and insulators.

9. Inspect and replace front suspension system leaf spring(s), leaf spring insulators (silencers), shackles, brackets, bushings, center pins/bolts, and mounts.

Removing and replacing front leaf springs are basic mechanical repair operations. Generally, leaf springs are replaced only when a leaf is broken or when ride height is below specification. To find worn leaf springs, ride height can be measured and compared to the vehicle manufacturer's specifications.

A leaf spring is mounted with a rubber bushing and bolt through the eye at one end and by rubber bushings and bolts on shackles at the other end. The shackles at one end of the spring let the spring length change as it flexes. If the spring were mounted directly to the frame at both ends, it would bind and eventually break.

Inspect shackles and bolts for damage and excessive wear. Inspect rubber bushings for wear, deterioration, and damage from grease and oil. Special removal and installation tools often make bushing replacement easier.

10. Inspect, replace, and adjust front suspension system torsion bars and mounts.

A torsion bar performs the same function as a coil or leaf spring. Its function is to support the vehicle weight and allow the wheels to follow the changes in the road surface and also absorb shock. The difference is, unlike a coil spring which compresses, the torsion bar uses a twisting action.

Removing and replacing torsion bars, like springs, are basic repair operations. Torsion bars are also generally replaced only when damaged. Unlike coil and leaf springs, torsion bar stiffness is adjustable on the vehicle, and this is what establishes the ride height of the vehicle.

One end of the torsion bar is splined or clamped to a suspension control arm. The other end is secured in a bracket on the chassis. The chassis end of the torsion bar has a short arm (sometimes called the "key") and an adjusting bolt to set the ride height and bar stiffness. Checking the ride height and adjusting it if necessary is a basic part of wheel alignment service. Car manufacturers' ride height specifications and measurement points vary, so you should check the manufacturer's instructions and specifications for this procedure. Usually, the final adjustment should be made in an upward direction. This helps to ensure that the vehicle will not settle to a lower than specified ride height when in operation.

Some vehicles are equipped with torsion bars mounted transversely.

If the torsion bars are removed, make sure they are reinstalled on their original side. When replacing torsion bars, check for indicating marks: left and right and/or front and rear. It is a good idea to lubricate the ends of the torsion bar with an anti-seize compound to prevent squeaks as the suspension travels through jounce and rebound.

11. Inspect and replace front stabilizer bar (sway bar) bushings, brackets, and links.

Stabilizer bars—also called antiroll bars or sway bars—minimize body roll, or sway, during cornering. Stabilizer bars do not affect spring stiffness or vehicle spring rate, ride height, or shock absorber action. A stabilizer bar is mounted in brackets with bushings on the car underbody or frame. Links attach each end of the bar to the front or rear control arms or axle housing. During cornering, the bar and its links transfer vehicle loads from the inside to the outside of the suspension. This reduces the tendency of the outside suspension to lift and thus reduces body roll.

The rubber bushings on stabilizer bars and links tend to deteriorate over time and can also be damaged by grease and oil. Worn or damaged bushings should be replaced. Mounting bolts and link bolts may become loose and occasionally break. These should be tightened or replaced as necessary.

12. Inspect and replace front strut cartridge or assembly.

A new cartridge may be installed in some front struts with the strut installed in the vehicle. Other struts must be removed to allow cartridge installation. Prior to strut removal from the vehicle, the upper strut mounting nuts and the strut-to-steering knuckle bolts must be removed. A spring compressor must be used to compress the spring before the spring is removed from the strut.

MacPherson struts not only are suspension parts, but also serve as shock absorbers and help control vehicle bounce. When you replace just the cartridge and not the outside housing, oil is left in the old housing to help transfer heat.

13. Inspect and replace front strut bearing(s) and mount(s).

A defective upper strut mount may result in strut chatter while cornering, poor steering wheel return, and improper camber or caster angles on the front suspension.

The caster and camber adjuster plates would make noise if someone had left the bolts loose. The bearings and support plates support the weight of the front of the chassis, the engine, and the transaxle, and have to withstand the weight-shifting forces of braking, and the rotating forces of steering the vehicle.

When the steering wheel wants to return to a position other than center, this is known as memory steer. Memory steering occurs when a steering component or bushing binds and prevents the steering gear from smoothly rotating back to center. That is why it is important that all steering components be tightened in their normal resting positions. Possible causes for memory steering are binding upper strut mounts, steering gear, linkage, ball joints, tie rods, and idler arm. Removing the steering linkage from the steering knuckles can help in isolating a binding component.

Sometimes the control valve in the steering gear fails and bypasses fluid into one side or the other of the boost cylinder piston, causing the steering to want to turn itself to one side. To check this condition, raise the vehicle off the ground and start the engine. If the wheels want to turn to one side with the engine running, suspect a faulty control valve or steering gear.

14. Inspect and replace shock absorbers, mounts, and bushings.

The function of the *shock absorber* is to control and dampen spring oscillations. Since shock absorbers are sealed units, no servicing is required. Inspect the shock bushings and mounts for wear. On most designs, the shock absorber will have to be replaced if the shock bushings are worn. Check for broken mounts and for physical damage to the shock.

When one side of the bumper is pushed downward with considerable force and then released, the bumper should only complete one free upward bounce if the shock absorber or strut is satisfactory. More than one free upward bounce indicates defective shock absorbers, loose shock absorber mountings, or defective struts.

Shocks should be inspected for oil leakage. If oil is dripping from the shock, replace it. The procedure to remove a front shock absorber:

1. Lift the vehicle on a hoist and support the suspension on safety stands so the shock absorbers are not fully extended.
2. Disconnect the upper shock mounting nut and grommet.
3. Remove the lower shock mounting nut or bolts.
4. Remove the shock absorber.

15. Diagnose and service front wheel bearings/hub assemblies.

The front wheel bearings perform a major role in effective brake function and steering performance. There are two different types of wheel bearing designs used on the front of modern automobiles: the adjustable tapered roller bearing and the non-adjustable sealed roller or ball bearing.

Worn wheel bearings can cause a growling noise and vibration when the vehicle is driven. Checking for loose or worn wheel bearings should always be part of a complete steering/suspension inspection. A worn or improper wheel bearing adjustment can cause poor brake performance, poor steering, and rapid wheel bearing wear.

Tapered wheel bearings can be disassembled, cleaned, re-greased, and adjusted. Sealed roller or ball bearing wheel bearings cannot be serviced and are replaced as a unit.

Tapered roller bearings are generally used on non-drive axles. The wheel bearings are mounted between a hub and a fixed spindle. To gain access to tapered wheel bearings, remove the wheel, brake rotor or drum, dust cap, cotter pin, spindle nut, washer, and remove the outer wheel bearing. Remove the hub/rotor or hub/drum assembly. Pry the wheel bearing seal and remove the inner wheel bearing. Thoroughly clean the wheel bearing, hub assembly, spindle shaft, and races.

Carefully inspect the wheel bearings, spindle, and races for signs of wear. Discard bearings showing any signs of wear, chipping, galling, or discoloration from overheating. Repack the wheel bearings with high-temperature grease. Never repack a wheel bearing without first removing all of the old grease. Insert the inner bearing into the hub and lightly lubricate the new wheel seal with grease. Tap the seal in place using a seal driver. Carefully install the hub/rotor or hub/drum assembly onto the spindle. Install the outer wheel bearing, washer, and spindle nut.

If the wheel bearings need to be replaced, it will be necessary to replace the bearing races as well. Remove the bearing race from the hub using a bearing race remover. If using a drift punch, tap the races a little at a time, moving the punch around the race to avoid cocking. Use a soft steel drift, never a hardened punch.

It is very important to properly adjust tapered wheel bearings. Always check with the manufacturer's recommended procedure for the specific vehicle being serviced. There are two widely used methods for adjusting wheel bearings: the torque wrench method and the dial indicator method.

In the torque wrench method, rotate the wheel in the direction of tightening while the spindled nut is tightened to the specified torque. This initial torque setting seats the bearings in the races. The nut is then loosened until it can be rotated by hand. The nut is then re-torqued to a lower specified value. Back the nut, if necessary, to install the cotter pin, and lightly tap on the dust cap.

To adjust the wheel bearings using the dial indicator method, start by tightening the spindle nut while spinning the wheel, to fully seat the bearings. Loosen the spindle nut until it can be rotated by hand. Mount a dial indicator so the indicator point makes contact to the machined outside face of the hub. Firmly grasp the sides of the rotor or tire and pull in and out. Adjust the spindle nut until the end play is within manufacturer's specification. Typical end play range is from 0.001 inch to 0.005 inch (0.025 mm to 0.125 mm). Install the cotter pin and dust cap. (A new cotter pin is typically recommended.)

Sealed ball and roller bearings are not serviceable and must be replaced when they are worn or defective or have damaged grease seals. Some designs of sealed bearings are simply bolted into the steering knuckle; others are pushed into the steering knuckle. A pushed in bearing often involves pressing the bearing from the hub of the spindle or knuckle assembly (there are special tools available so this procedure can be accomplished on the vehicle). Carefully inspect the bearing, hub, and spindle assembly for wear. Press the new bearing into the spindle/knuckle assembly and torque the axle nut following the manufacturer's procedure.

Constant velocity
universal joint (4)

Steering
knuckle (3)

Wheel
bearing (2)

Hub (1)

Knuckle ring (5)

Snap ring (6)

Washer (7)

Hub nut (8)

1. Cubo
2. Rodamiento de rueda
3. Mangueta de la dirección
4. Junta homocinética (Unión universal de Velocidad Constante – CV)
5. Aro / Anillo de la mangueta
6. Anillo de presión
7. Arandela / Roldana
8. Tuerca del cubo

Front wheel drive wheel bearing

16. Diagnose, inspect, adjust, repair or replace components (including sensors, switches, and actuators, and control units) of electronically controlled suspension systems (including primary and supplemental air suspension and ride control systems).

Automatic level control systems are designed to maintain correct vehicle ride height under different load changes. Level control systems use air pressure that is pumped into air shocks or air springs in response to different load changes. Air pressure is developed by an electrically operated pump. Some systems use a dryer assembly to absorb moisture from the system. This reduces the chance of corrosion damage to internal components. Some vehicles have a manual shutoff switch to disable the automatic level control system. On these vehicles, the switch must be turned off whenever the vehicle is raised off the ground from the frame with the wheels hanging.

Once the air suspension system has been shut down for an hour, it becomes inactive. If there are leaks, the vehicle ride height will decrease when the vehicle is parked and not in use. It is normal for an air suspension system to drop a little overnight, especially when there is a significant temperature change. If the system is functioning properly, the vehicle will level itself soon after startup. Most air suspension systems, both primary and supplemental, are automatic and have height level sensors, air control solenoids, relays, an electric air pump, and a module to make the system work. Most modern systems will store fault codes to help with diagnosis, and some systems have a function test that allows each corner of the vehicle to be raised and lowered to verify operation.

Use a scan tool for diagnosing the electronic suspension. Refer to the scan tool manufacturer's instructions for specific information.

An inspection of an automatic level control system should include the compressor, height sensors, hoses, hose connections, air shocks (or air struts), electrical connectors, relays, solenoids, wire harness, electrical components, dryer, and pressure regulator. Consult the factory manuals for detailed information on various systems.

The computerized ride control system is another automatic suspension design. These suspensions are computer controlled and can adapt to different road conditions and driving situations. Hydraulic pressure is redirected and controlled by the use of actuators within the shock absorber or strut. In this way, the suspension can be altered from a soft ride to a stiffer ride. Some systems will allow the driver to select the type of ride desired for the particular road conditions. For example, a firmer, more controlled setting would be more desirable on a winding road. For highway cruising, a softer mode would be more applicable. On many systems, the computer may override any pre-set modes. If the vehicle is under a hard braking situation, the controller will stiffen up the front shocks to help maintain vehicle control. The same will be true under heavy acceleration. The controller will stiffen the rear shocks to minimize rear-end squatting. When the driver turns hard into a turn, the controller will stiffen the outside shocks and reduce body roll.

Typical computerized ride control systems incorporate a control module, brake sensors, steering sensors, acceleration sensors, mode select switch, actuators, and height sensors. Diagnosing a computerized suspension system varies for different manufacturers. It will be necessary to follow the diagnostic procedures in the appropriate service manual. If the computer recognizes a problem within the system, it will alert the driver by illuminating a warning light on the dashboard. A trouble code may be stored in the memory of the computer, which will aid in the diagnostic process.

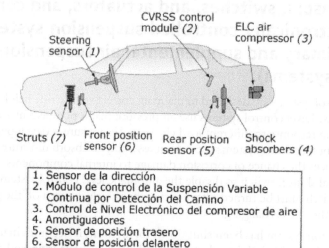

Computer controlled suspension

17. Inspect and repair front cradle (crossmember/subframe) mountings, bushings, brackets, and bolts.

On unibody vehicles, a subframe is used to help support and locate the drive train. On vehicles with a frame and on some unibody vehicles, a crossmember is used to support the engine and/or transmission. Proper alignment of the drive train is critical to the handling of the vehicle and the operation of many systems of the vehicle. As an example, if the cross-member or subframe is not secured to the vehicle or if the mounting's bushings are worn, the driver may experience shifting problems due to the misalignment of the shift linkage.

When you service these units, the weight of the engine and/or transmission must be relieved before you perform any service. This is often done by securing the engine on a hoist.

2. Rear Suspensions

1. Diagnose rear suspension system noises, handling, ride height, and ride quality concerns; determine needed repairs.

A squeaking noise in the rear suspension may be caused by suspension bushings, defective struts or shock absorbers, or broken springs or spring insulators. Harsh riding may be caused by reduced rear suspension ride height and defective struts or shock absorbers.

Excessive rear suspension oscillations may be caused by defective struts. Weak coil springs cause harsh riding and reduced ride height. Broken springs or spring insulators cause a rattling noise while driving on irregular road surfaces. Worn out struts or shock absorbers result in chassis oscillation and harsh riding.

A broken spring leaf will cause the vehicle to lean toward the broken side. Missing insulators will cause creaking and squeaking noises, not rattles, as the suspension moves up and down. Worn shackle bushings or worn shackles will cause rattles when the vehicle is driven over road irregularities at low speeds. Broken center bolts will allow one side of the axle or housing to move forward or rearward.

Sway bars are not likely to cause vibrations. Coil springs with a high-load rating could cause the vehicle to be too high in the rear.

If the rear strut cartridge is weak, the vehicle will bounce more than normal in the rear, but it should not hit bottom going over speed bumps at low speeds. If the rear springs are weak, the chassis will hit bottom easily.

2. Inspect and replace rear suspension system coil springs and spring insulators (silencers).

When replacing rear coil springs, the old spring ends should be matched with the new springs. Matching the spring ends will ensure that the springs are installed correctly. Linear rate springs or variable-rate springs may be used. Linear rate springs have equal spacing between the coils and are available as heavy-duty springs for most applications. Variable rate springs typically have coils spaced closer together at the top, with more space between the coils at the bottom of the spring. Variable rate springs provide automatic load adjustment while maintaining vehicle height.

With the springs removed, inspect the spring insulators and spring seats. Worn or cracked insulators should be replaced. Check for rusted or damaged spring seats. When installing the coil springs, make sure the spring is aligned properly in the spring seat. After the springs are installed, check vehicle ride height and road test the vehicle.

It may be necessary to compress the springs and to disconnect the shock absorber and other components in order to remove the springs.

3. Inspect and replace rear suspension system lateral links/arms (track bars), control (trailing) arms, stabilizer bars (sway bars), bushings, and mounts.

Different rear suspension designs utilize a variety of components. Generally, rear suspension designs use coil springs, leaf springs, or MacPherson struts. When coil springs are used, a system of control arms, links, and/or track bars are used to maintain stability and alignment of the rear axle assembly. The control arms, lateral link, and track bar pivot points are insulated by rubber bushings. Inspect all bushings for wear and check all components for damage, twisting, or bending.

When replacing suspension bushings, it is important that the weight of the vehicle be on the bushings before you tighten the bolts. All bushings should be tightened at their normal stand-ing ride height. Damage to the bushings will occur if they are not tightened in their normal resting position. This procedure also applies to track bars and lateral arms. It may be necessary to use a press or suitable tool in order to install rear suspension bushings. Rubber bushings should be lubricated only with an approved lubricant. Never lubricate with oil or grease.

Some rear suspensions have adjustable driveline (pinion) angles. Before installing control arms, check with the factory manual. On some vehicles, an eccentric washer on the control arm adjusts the driveline (pinion) angle.

The rear stabilizer bar, or sway bar, helps to control body roll on turns. The twisting action of the bar counteracts body sway on turns and holds the vehicle closer to a level riding position. Inspect the connecting links, link bushings, and mounting bushings for wear. If the sway bar is bent or damaged, replace it. As with other bushings, tighten only at their normal resting position.

4. Inspect and replace rear suspension system leaf spring(s), leaf spring insulators (silencers), shackles, brackets, bushings, center pins/bolts, and mounts.

Inspect the leaf springs, bushings, insulators, shackles, and mounts for wear and damage. The most common failure point on a rear leaf suspension is the spring bushings. Check the vehicle ride height to determine the condition of the springs. Leaf springs may sag or break over time. Carefully inspect each leaf in a multiple-leaf design for broken leaves, worn insulators, and broken spring retaining clips. Inspect the frame for rust and corrosion at the locations where the spring shackles and mounts are attached. If the springs are not aligned properly with the rear axle or if the rear axle has shifted, check for a broken spring center bolt.

To remove the rear springs, raise the vehicle off the ground, support the rear axle assembly, and remove the wheels. Disconnect the shock absorbers; unbolt the rear shackles and the front mount from the frame. Remove the attaching U-bolts from the axle and remove the springs from the vehicle. Replace all worn bushings and other components, as required. After reinstalling the springs, tighten all bushings with the vehicle weight on the springs and at normal ride height.

5. Inspect and replace rear rebound and jounce bumpers.

The rear rebound and jounce bumpers serve the same purpose as in the front. When the wheel hits a bump, the control arms pivot upward causing the spring and shock to compress. Rubber bumpers cushion the blow, should the control arms reach their limit of travel. Inspect the rebound or jounce bumpers for wear and cracks. In some cases, the bumper may actually be missing. Worn bumpers can be caused by low ride height settings, worn springs, and/or worn shocks or struts.

6. Inspect and replace rear strut cartridge or assembly, and upper mount assembly.

Rear struts are serviced similarly to front struts, except that rear struts do not have a steering knuckle. The coil spring on the strut must be compressed to separate it from the strut assembly. Some struts can be serviced by replacing an internal cartridge that contains the shock absorber. Others require replacement of the entire strut.

When reassembling the strut, be sure that the spring is seated securely in its mounting brackets. Inspect the upper mounting location on the car body. Replace any worn or damaged fasteners or other parts. If the body structure is damaged, more extensive repairs will be required.

7. Inspect non-independent rear axle assembly for damage and misalignment.

A non-independent rear axle may be checked for bending, warpage, and misalignment by measuring the rear-wheel tracking. This operation may be performed with a track bar or computer wheel aligner with four-wheel capabilities. A track bar measures the position of the rear wheels in relation to the front wheels. A computerized wheel aligner displays the thrust angle, which is the difference between the vehicle thrust line and geometric centerline of the vehicle. Rear axle offset can cause a steering wheel that is not centered and may cause steering pull.

On vehicles with rear leaf springs, check the center bolt. If the center bolt is broken, the rear axle assembly may shift, causing misalignment. This would show up while you perform a wheel alignment. The thrust angle and rear toe will not be correct. (Thrust angle and toe will be covered in the wheel alignment tasks).

8. Inspect and replace rear ball joints and tie rod/toe link assemblies.

Some ball joints may have a wear indicator. In these ball joints, the shoulder of the grease fitting must extend a specific distance from the ball joint housing. If this distance is less than specified, the ball joint must be replaced. This measurement is made while the ball joint is loaded. A worn ball joint may cause improper positioning of the lower end of the rear knuckle, wheel hub, and wheel. This action may result in improper rear-wheel camber.

If a rear-wheel tie rod is longer than specified, the rear-wheel toe-out will be out of alignment. The length of the tie rod determines the rear-wheel toe setting.

Rear load-carrying and non-load-carrying ball joints are tested like the front ball joints. Rear tie rod ends are checked the same as front tie rod ends. The rear ball joints and rear tie rod ends usually last much longer than the front, because the rear does not rotate and these components carry much less weight. Some rear ball joints and rear tie rod ends have to be lubricated.

9. Inspect and replace rear knuckle/spindle assembly.

The steering knuckle, or wheel spindle, is the mounting point for the wheel and brake assemblies. The wheel rotates on the spindle shaft via a set of bearings. The steering knuckle/spindle is held in place by control arms and/or the suspension strut. To replace a knuckle/spindle, remove the wheel assembly and disconnect the ball joints, control arms, steering linkage, springs, and/or strut assembly from the spindle assembly. After reinstallation, the wheels must be aligned.

10. Inspect and replace shock absorbers, mounts, and bushings.

The function of the *shock absorber* is to control and dampen spring oscillations. Since shock absorbers are sealed units, no servicing is required. Inspect the shock bushings and mounts for wear. On most designs, the shock absorber will have to be replaced if the shock bushings are worn. Check for broken mounts and for physical damage to the shock.

When one side of the bumper is pushed downward with considerable force and then released, the bumper should only complete one free upward bounce if the shock absorber or strut is satisfactory. More than one free upward bounce indicates defective shock absorbers, loose shock absorber mountings, or defective struts.

Shocks should be inspected for oil leakage. If oil is dripping from the shock, replace it. The procedure to remove a rear shock absorber:

1. Lift the vehicle on a hoist and support the suspension on safety stands so the shock absorbers are not fully extended.
2. Disconnect the upper shock mounting nut and grommet.
3. Remove the lower shock mounting nut or bolts.
4. Remove the shock absorber.

11. Diagnose and service rear wheel bearings/ hub assemblies.

The rear wheel bearings (like the front wheel bearings) perform a major role in effective brake function and steering performance. There are three different types of wheel bearing designs used on the rear of modern automobiles: the adjustable tapered roller bearing, the non-adjustable sealed roller or ball bearing, and the wheel bearing installed in the non-independent rear drive axle.

Worn wheel bearings can cause a growling noise and vibration when the vehicle is driven. Checking for loose or worn wheel bearings should always be part of a complete steering/ suspension inspection. A worn or improper wheel bearing adjustment can cause poor brake performance, poor steering, and rapid wheel bearing wear.

Tapered wheel bearings can be disassembled, cleaned, re-greased, and adjusted. Sealed roller or ball bearing wheel bearings cannot be serviced and are replaced as a unit.

On the rear suspension, tapered roller bearings can be used on non-drive axles as well as on full floating rear axles. The wheel bearings are mounted between a hub and a fixed spindle (axle housing). To gain access to tapered wheel bearings, remove the wheel, brake rotor or drum, dust cap (or axle), cotter pin, and spindle nut (axle nut), washer, and remove the outer wheel bearing. Remove the hub/rotor or hub/drum assembly. Pry the wheel bearing seal and remove the inner wheel bearing. Thoroughly clean the wheel bearing, hub assembly, spindle shaft, and races.

Carefully inspect the wheel bearings, spindle, and races for signs of wear. Discard bearings showing any signs of wear, chipping, galling, or discoloration from overheating. Some of these bearings will be grease packed, others will be lubricated by the lubricating oil in the drive axle. On grease packed bearings, repack the wheel bearings with high-temperature grease. Never repack a wheel bearing without first removing all of the old grease. Insert the inner bearing into the hub and lightly lubricate the new wheel seal with grease. Tap the seal in place using a seal driver. Carefully install the hub/rotor or hub/drum assembly onto the spindle. Install the outer wheel bearing, washer, and spindle nut.

If the wheel bearings need to be replaced, it will be necessary to replace the bearing races as well. Remove the bearing race from the hub using a bearing race remover. If using a drift punch, tap the races a little at a time, moving the punch around the race to avoid cocking. Use a soft steel drift, never a hardened punch.

It is very important to properly adjust tapered wheel bearings. Always check with the manufacturer's recommended procedure for the specific vehicle being serviced. There are two widely used methods for adjusting wheel bearings: the torque wrench method and the dial indicator method.

In the torque wrench method, rotate the wheel in the direction of tightening while the spindled nut is tightened to the specified torque. This initial torque setting seats the bearings in the races. The nut is then loosened until it can be rotated by hand. The nut is then re-torqued to a lower specified value. Back the nut, if necessary, to install the cotter pin, and lightly tap on the dust cap. A new cotter pin is recommended.

To adjust the wheel bearings using the dial indicator method, start by tightening the spindle nut while spinning the wheel, to fully seat the bearings. Loosen the spindle nut until it can be rotated by hand. Mount a dial indicator so the indicator point makes contact to the machined outside face of the hub. Firmly grasp the sides of the rotor or tire and pull in and out. Adjust the spindle nut until the end play is within manufacturer's specification. Typical end play range is from 0.001 inch to 0.005 inch (0.025 mm to 0.125 mm). Install the cotter pin and dust cap. A new cotter pin is recommended.

Sealed ball and roller bearings are not serviceable and must be replaced when they are worn or defective or have damaged grease seals. Some designs of sealed bearings are simply bolted into the steering knuckle; others are pushed into the steering knuckle. A pushed in bearing often involves pressing the bearing from the hub of the spindle or knuckle assembly (there are special tools available so this procedure can be accomplished on the vehicle). Carefully inspect the bearing, hub, and spindle assembly for wear. Press the new bearing into the spindle/knuckle assembly and torque the axle nut following the manufacturer's procedure.

12. Diagnose, inspect, adjust, repair or replace components (including sensors, switches, and actuators, and control units) of electronically controlled suspension systems (including primary and supplemental air suspension and ride control systems).

Automatic level control systems are designed to maintain correct vehicle ride height under different load changes. Level control systems use air pressure that is pumped into air shocks or air springs in response to different load changes. Air pressure is developed by an electrically operated pump. Some systems use a dryer assembly to absorb moisture from the system. This reduces the chance of corrosion damage to internal components. Some vehicles have a manual shutoff switch to disable the automatic level control system. On these vehicles, the switch must be turned off whenever the vehicle is raised off the ground from the frame with the wheels hanging.

Once the air suspension system has been shut down for an hour, it becomes inactive. If there are leaks, the vehicle ride height will decrease when the vehicle is parked and not in use. It is normal for an air suspension system to drop a little overnight, especially when there is a significant temperature change. If the system is functioning properly, the vehicle will level itself soon after startup. Most air suspension systems, both primary and supplemental, are automatic and have height level sensors, air control solenoids, relays, an electric air pump, and a module to make the system work. Most modern systems will store fault codes to help with diagnosis, and some systems have a function test that allows each corner of the vehicle to be raised and lowered to verify operation.

Use a scan tool for diagnosing the electronic suspension. Refer to the scan tool manufacturer's instructions for specific information.

An inspection of an automatic level control system should include the compressor, height sensors, hoses, hose connections, air shocks (or air struts), electrical connectors, relays, solenoids, wire harness, electrical components, dryer, and pressure regulator. Consult the factory manuals for detailed information on various systems.

The computerized ride control system is another automatic suspension design. These suspensions are computer controlled and can adapt to different road conditions and driving situations. Hydraulic pressure is redirected and controlled by the use of actuators within the shock absorber or strut. In this way, the suspension can be altered from a soft ride to a stiffer ride. Some systems will allow the driver to select the type of ride desired for the particular road conditions. For example, a firmer, more controlled setting would be more desirable on a winding road. For highway cruising, a softer mode would be more applicable. On many systems, the computer may override any pre-set modes. If the vehicle is under a hard braking situation, the controller will stiffen up the front shocks to help maintain vehicle control. The same will be true under heavy acceleration. The controller will stiffen the rear shocks to minimize rear-end squatting. When the driver turns hard into a turn, the controller will stiffen the outside shocks and reduce body roll.

Typical computerized ride control systems incorporate a control module, brake sensors, steering sensors, acceleration sensors, mode select switch, actuators, and height sensors. Diagnosing a computerized suspension system varies for different manufacturers. It will be necessary to follow the diagnostic procedures in the appropriate service manual. If the computer recognizes a problem within the system, it will alert the driver by illuminating a warning light on the dashboard. A trouble code may be stored in the memory of the computer, which will aid in the diagnostic process.

13. Inspect and repair rear cradle (crossmember/subframe) mountings, bushings, brackets, and bolts.

On unibody vehicles, a subframe is used to help support and locate the drive train. On vehicles with a frame and on some unibody vehicles, a crossmember is used to support the engine and/or transmission. Proper alignment of the drive train is critical to the handling of the vehicle and the operation of many systems of the vehicle. As an example, if the cross-member or subframe is not secured to the vehicle or if the mounting's bushings are worn, the driver may experience shifting problems due to the misalignment of the shift linkage.

When you service these units, the weight of the engine and/or transmission must be relieved before you perform any service. This is often done by securing the engine on a hoist.

C. Wheel Alignment Diagnosis, Adjustment, and Repair (11 Questions)

1. Diagnose vehicle wander, drift, pull, hard steering, bump steer (toe curve), memory steer, torque steer, and steering return concerns; determine needed repairs.

Wheel alignment is the process of measuring and correcting steering and suspension angles. Proper wheel alignment is desired in order to control the vehicle in a safe and predictable manner. Incorrect wheel alignment can cause hard steering, premature tire wear, pulling to one side, wandering, and decreased steering performance. Before a wheel alignment is performed, a complete inspection should be done on the suspension and steering system. Worn steering components or bent suspension parts will prevent the vehicle from being aligned properly. Sagging springs, broken springs, worn wheel bearings, or loose wheel bearings will also affect the wheel alignment.

The first step in the alignment process starts with a road test. Take notice of pulling, wandering, wheel shimmy, and any other steering problems. Does the problem seem to change during braking? Does the vehicle track well in a straight line but feel "loose" when turning? Does it turn better in one direction than the other? Are there any noises from the suspension system? These are all items you should be observing when you are driving the vehicle. Develop a test route and drive it each time to duplicate all kinds of driving conditions and turns.

Tire pressure must be checked and adjusted while the vehicle is at the correct ride height. Setting the caster, camber, and toe without correcting ride height may not correct tire wear and handling problems. Incorrect ride height in front-wheel drive vehicles may cause vibrations, especially during acceleration. All steering linkage parts should have zero freeplay and should be replaced if any looseness is felt or measured during inspection.

In all vehicles, the rear alignment must be correct before any adjustments are made to the front. For vehicles with rear-wheel steering, the rear steering system must remain in its centered position while the front adjustments are being checked and adjusted.

Memory steering occurs when a steering component or bushing binds and prevents the steering gear from smoothly rotating back to center. That is why it is important that all steering components be tightened in their normal resting position. Possible causes for memory steering are binding upper strut mounts, steering gear, linkage, ball joints, tie rods, and idler arm. Removing the steering linkage from the steering knuckles can help in isolating a binding component.

Sometimes the control valve in the steering gear fails and bypasses fluid into one side or the other of the boost cylinder piston, causing the steering to want to turn itself to one side. To check this condition, raise the vehicle off the ground and start the engine. If the wheels want to turn to one side with the engine running, suspect a faulty control valve or steering gear.

2. Measure vehicle ride height; determine needed repairs.

Vehicle ride height is an important specification and must be checked before the alignment is performed. Ride height is usually adjustable on a vehicle with torsion bars.

Ride height can vary significantly on a single model of a light truck with various spring and wheel-and-tire combinations. As ride height varies, so does the front camber angle. Many trucks have different camber specifications for different ride heights. Most truck manufacturers publish tables of varying ride height specifications, which should be checked during the alignment operation.

Ride height measurement points vary from one vehicle to another. Some are measured between the lower control arm and the ground. Others are measured between a point on the fenderwell or under body and ground. Always verify the vehicle manufacturer's measurement points, as well as the specifications. On some vehicles, typically light trucks, the vehicle manufacturer provides alignment specifications that can compensate for changes in ride height due to spring sag. In cases such as this, a frame angle measurement may be required to be taken and input into the alignment instrumentation. Follow your alignment instrumentation manufacturer's instructions.

If ride height is out of limits on a vehicle with coil or leaf springs, the springs or other suspension parts may require replacement.

Most vehicle manufacturers call for wheel alignment adjustments with the vehicle unloaded and at a specified ride height. Some carmakers, however, specify precise weight loads to be placed in a car during alignment. Trucks are often aligned with specified loads.

3. Measure front and rear wheel camber; determine needed repairs.

Camber is the outward or inward tilt of the wheel as viewed from the top of the tire. The more inward the top of the wheel is from true vertical, the more negative the camber. The more outward the wheel is from true vertical, the more positive the camber. Incorrect camber may cause increased road shock and pulling to one side. Camber will usually pull to the most positive side. Also, incorrect camber can lead to rapid tire wear. Camber should not vary more than 0.5 degrees from side to side. This is typically referred to as "cross camber." A vehicle will tend to pull to the side that has the most positive camber.

If camber is not adjustable, inspect the struts, suspension, and steering system for bent components. Replace damaged or bent components as needed, and recheck camber angles.

4. Adjust front and/or rear wheel camber on suspension systems with a camber adjustment.

Vehicle manufacturers provide many different ways to adjust front and rear camber:

1. Shims may be placed between various suspension components and the frame. Shims are usually used between control arm pivot shafts and their mounting brackets on the frame.
2. An eccentric cam lobe may be turned to move the control arm pivot point inward or outward on the chassis.
3. Adjust sleeves on control arm linkage.
4. Move strut mounts.

Vehicle manufacturers publish wheel alignment specifications annually, and most computerized alignment equipment contains an onboard database of specifications and adjustment instructions.

Some vehicles have slightly more positive camber on the left front wheel than on the right to minimize vehicle pull caused by the crown of the road. More often, however, road crown compensation is done with slightly different caster angles.

Front-wheel camber and caster are adjusted simultaneously on some vehicles that provide adjustment. Before adjusting camber and caster, jounce the vehicle to relieve any binding or stress on suspension parts and let it settle at its normal ride height. When either angle is adjusted, the other should be checked because changing one will affect the other. The front-wheel toe angle is adjusted after caster and camber adjustments are done.

5. Measure caster; determine needed repairs.

Caster is the tilting of the spindle support centerline from true vertical. Measuring the position of the lower ball joint in relation to the upper ball joint, or the top of the MacPherson strut, determines the caster angle. The spindle support centerline is an imaginary line drawn through the center of the upper ball joint and lower ball joint, or the ball joint and the center of the upper strut mount on a MacPherson strut design. If the lower ball joint is positioned more forward than the upper ball joint, the caster angle is positive. If the lower ball joint is set more rearward than the upper ball joint, the caster angle is negative. On a MacPherson strut design, if the ball joint is positioned more forward than the center of the upper strut mount, the caster angle is positive.

Caster helps improve steering effort, high-speed stability, and steering wheel returnability. Correct caster angles will keep the vehicle traveling in a straight line going forward. On most vehicles, caster is set positive and is usually set at the same degree on both sides or with slightly more on the right side of the vehicle to compensate for the drainage crown built into most roads.

Too much positive caster will cause hard (heavy) steering, road shock, and wheel shimmy. Caster set too negative will cause wandering. Caster should not usually vary more than 0.5 degrees from one side to the other. This is typically referred to as "cross caster." If caster is not within specification and no adjustments are provided, inspect the suspension and steering for bent or damaged components. Also, if the vehicle was involved in a collision, check for frame damage. A vehicle will tend to pull to the side that has the least positive caster. Incorrect caster will not cause tire wear.

6. Adjust caster on suspension systems with a caster adjustment.

For vehicles with adjustable caster, consult the service manual for specifications and procedures. Before making caster adjustments, perform a complete steering and suspension inspection. On some vehicles, the caster is adjusted by moving the lower strut rod. By lengthening or shorting the lower strut rod, you can bring caster into specification. On some MacPherson strut vehicles, loosening the upper strut mount and sliding it forward or backward adjusts caster. Vehicles with short/long arm suspension (SLA) often provide caster adjustments. Shimming or sliding the upper control arm will adjust caster. Some upper control arms are designed with eccentric cams, which are rotated to adjust caster. Adjusting caster by moving the upper control arm can also incorporate camber adjustment. Some light trucks have an offset upper ball joint bushing, which will provide caster adjustment. Bushings come in different sizes, which correspond to different degree changes in caster. For example: If you want to change caster 0.5 degrees positive and the bushing in the original upper ball joint bushing is 0 degrees, remove the old bushing and install a 0.5 degrees bushing. Make sure the bushing is inserted in the correct position to achieve positive 0.5 degrees. Consult the service manual for procedure on removing and installing the bushing. There are other methods of adjusting caster. Check the service manual for a specific vehicle.

7. Measure and adjust front wheel toe.

Setting front toe is performed after camber, caster (front and rear), and rear toe have been checked and/or adjusted. Turn the steering wheel from side to side, center it, and then lock it in place using the appropriate tool. If the vehicle is equipped with power steering, have the engine running when centering the steering wheel. Measure front toe and check factory specifications. Front toe is adjusted by turning the tie rods until correct toe is attained. On conventional steering systems, loosen the tie rod sleeve clamps and turn the sleeve to change toe. For rack and pinion steering, loosen the locknuts and rotate the inner tie rods. Some vehicles have only one adjusting tie rod. On these vehicles, the steering wheel is not locked in place. Follow recommended procedures for adjusting toe and centering the steering wheel on these vehicles. On vehicles with adjusting sleeves, make sure the adjusting clamps are re-tightened in the correct position, so as not to interfere with other steering linkage or other components when the wheels are turned. After toe has been adjusted, the vehicle should be road tested. The steering wheel should be straight with the wheels in the straight-ahead position.

8. Center the steering wheel.

If front toe is set correctly, the steering wheel should be centered correctly. If the steering wheel is not straight, it may be necessary to recheck toe. Remember to bounce the suspension and to idle vehicles with power steering after toe adjustments are made to be sure they are correct. Also, make sure the vehicle is tracking correctly. A misaligned rear suspension or twisted frame will affect steering wheel centering. If the steering is not straight, be sure to recheck the rear settings, particularly the thrust line.

9. Measure toe-out-on-turns (turning radius/angle); determine needed repairs.

The turning radius of the wheel on the inside of the turn is usually two degrees more than the turning radius on the outside of the turn. When the turning radius/angle is not within specifications, a steering arm may be bent.

Turn the front of the right wheel inward 20 degrees and read the turntable indicator on the left wheel. It should be slightly less than 20 degrees. Perform the same procedure by turning the front of the left wheel inward 20 degrees. Compare these readings and check the service manual for exact specifications. Turning radius angles (toe-out-on-turns) are built into the steering arm design. If the angles are not correct, check the steering arms, steering knuckles, and suspension for bent or damaged components. Incorrect toe-out-on-turns can cause the front tires to squeal when in a turn.

10. Measure SAI/KPI (steering axis inclination/king pin inclination); determine needed repairs.

Steering axis inclination (SAI) is not adjustable, but it should be measured and used to analyze handling problems or tire wear problems. A complete steering and suspension inspection along with a wheel alignment should be done first before checking steering angle inclination. SAI is the imaginary line formed by the inward tilt of the upper ball joint or strut mount. The angle is measured in degrees from true vertical. If SAI is incorrect, check for a bent component or frame.

11. Measure included angle; determine needed repairs.

The included angle is determined by adding the camber and steering angle inclination (SAI) together on any one wheel. A large difference from side to side is an indication of a bent suspension component or steering knuckle, or even frame damage. The included angle is not adjustable and is used to help diagnose steering or tire wear problems.

12. Measure rear wheel toe; determine needed repairs or adjustments.

Rear toe is usually adjusted by eccentric cams or threaded rods. On some models, rear toe is adjusted by loosening a lock-bolt and moving the suspension until the correct toe is attained. Shims installed between the rear axle flange and the wheel bearing can be used to adjust toe when no adjusters are provided. Check with shim manufacturers for specific procedures and application. On models with no adjustments, inspect the rear suspension for misaligned, bent, or damaged components.

On vehicles with four-wheel steering, check the appropriate service manual for specific procedures. Usually, a special tool is needed to lock the rear steering gear in place while rear toe and rear camber adjustments are made. The tool is removed when the front camber and caster are set, but reinstalled to adjust front toe.

13. Measure thrust angle; determine needed repairs or adjustments.

Imagine that the vehicle has an exact geometric centerline that runs through the center of it. Now imagine a live axle such as you would find in rear-wheel drive vehicles with both rear wheels running exactly parallel to the geometric centerline of the vehicle. This angle of the tires is the thrust angle.

As long as the angle is the same, or very close, to the same as the geometric centerline of the vehicle, it will track straight down the road. If the toe is pointed out on one side and in on the other, the vehicle will go in that direction when moving. You have probably seen a vehicle that appears to be going down the road with the rear out to one side or the other of the front wheels. This is often called "dog-tracking" and is really a thrust angle problem. Remember that in a rear-wheel drive application, the rear wheels steer the car and the front wheels correct for the rear. In front-wheel drive applications, a vehicle can still "dog-track," but it will not normally be noticed by the driver until the rear tires start to wear out. In most situations, it will be one tire that wears more than the other.

Thrust line issues can be caused by only one rear tire being significantly toed in or out. The triangulation of both rear toe readings is used to find the thrust line. Take the point of the triangle, whether it is ahead or behind the vehicle, and where it points away from the vehicle centerline is the thrust angle. Causes of thrust angle problems on any vehicle equipped with rear independent suspension include: incorrect toe settings; bent rear components like knuckles, control arms, and radius arms; bent or shifted rear suspension cradles; or frame damage. Thrust angle problems on rear-wheel drive live axle vehicles may be caused by worn spring eye bushings, broken leaf springs or broken leaf spring center bolts, loose or damaged leaf spring U-bolts, worn control arm bushings, bent frame or suspension mounting points, or a bent axle housing.

14. Measure front wheelbase setback/offset; determine needed repairs or adjustments.

Setback occurs when one front wheel is rearward in relation to the opposite front wheel. Front wheelbase setback is usually caused by collision damage. In other cases, it is designed into the vehicle.

It is possible for caster and camber adjustments to be within specification while the setback is excessive. In that case, the vehicle will pull to the side with the most setback because that side has a shorter wheelbase.

15. Check front and/or rear cradle (crossmember/ subframe) alignment; determine needed repairs or adjustments.

The front cradle may be measured in various locations to verify a bent condition. Some cradles have an alignment hole that must be aligned with a matching hole in the chassis.

The subframe is a critical part of the vehicle suspension and affects the steering. It must be centered properly in its designated location before the mounting bolts are tightened. Vehicle collisions can cause damage to the subframe, which will show up as excessive setback. Subframes must be centered properly with the chassis before the bolts are torqued.

16. Perform electronic control module calibration/ recalibration; perform initialization or relearn procedure as required.

The electronic control module will need to be calibrated after replacement or recalibrated after wheel alignment. A typical procedure will have the technician first perform a complete wheel alignment, including measuring and adjusting where possible rear toe/ thrust line. After all alignment adjustments are completed and within factory specifications the technician will steer the wheels straight ahead using the alignment machine as a guide. Then connect a scan tool and allow the control module to relearn the steering wheel position sensor. This will allow the electronic power steering system, lane departure, and radar cruise control to work in conjunction with the new alignment settings.

17. Diagnose damaged component mounting locations which can cause vibration, steering, and wheel alignment problems in accordance with vehicle manufacturer/industry recommended procedures.

When a vehicle cannot be aligned to factory specifications or does not drive correctly after an alignment, the technician should inspect for hidden structural damage. While structural damage can result from improper securing onto a transport vehicle, it is most often caused by a vehicular accident. New components, body work, or repainting of vehicle areas can be an indication of a repair after an accident. Manufacturers provide specific data for their vehicles that are typically used by body shops to ensure the body is restored to its original dimensions. These measurements are taken from jig holes and other locations from one side of the vehicle to the other, then cross checked with diagonal measurements to ensure the vehicle is square. Specific dimensions that would affect alignment angles would include upper strut tower, control arm, and steering component mounting locations.

D. Wheel and Tire Diagnosis and Service (5 Questions)

1. Diagnose tire wear patterns; determine needed repairs.

Feathered tire wear may be caused by improper toe adjustment. Wear on one side of the tire tread usually indicates an improper camber setting. Cupped tread wear may indicate improper wheel balance, worn shocks, or worn suspension components.

Tires wear excessively because the tire tread is not contacting the road surface properly. Alignment adjustments correct most of that problem. Normally, vehicles are aligned with no extra loads in the cargo area or passenger compartment. There are special circumstances, such as overweight drivers or drivers who do not distribute cargo weight properly, that require that the vehicle be similarly loaded while the alignment is made.

2. Inspect tire condition, size, and application (load and speed ratings).

Tires play a vital role in braking performance, handling ability, and ride quality. Defective tires, improper inflation, or installing the wrong tires can have a negative effect on overall vehicle steering and handling.

Different vehicles with different suspensions are designed to be driven with a specific tire construction, tire size, and tire pressure. The load exerted on the tires of a pickup truck used for hauling heavy cargo is quite different from that of a small compact sedan. Using the wrong tires on a vehicle or not inflating the tires correctly can cause serious results such as tire damage and may affect steering ability. Always install the correct tire size and adjust pressure to specification. Tire load information labels can generally be found on the door pillar, driver's door, or inside the glove box. The label will have specific information on tire size, inflation pressure, and the maximum vehicle load.

All tires ratings are determined by a series of numbers located on the tire. The rating system identifies the size, wheel size, type of construction, speed rating, and load rating. A typical tire may have the following designation: P 225/70 H R 15. The letter P indicates that this tire is designed for passenger vehicles and some light trucks. Other designations are T for temporary use, LT for light truck, and C for large trucks or commercial. The number 225 indicates the width of the tire measured in millimeters. The higher the number, the wider the tire. The number 70 is the aspect ratio of the tire's height to its width. The higher the aspect ratio number the taller the tire. The letter H designates the speed rating. The rating specifies the maximum speed at which the tire can be driven safely. The rating range is from B (31 mph, 49 kph) to Z (over 149 mph, 240 kph). The letter R indicates a radial ply construction. Bias ply tires are marked B. The number 15 is the rim diameter in inches.

Tires are also graded for temperature resistance, traction, and tread wear. Temperature resistance is graded on three levels: A, B, and C. "A" provides the greatest resistance to heat; "C" provides the least. Traction is rated on four levels: AA, A, B, and C. "AA" will provide the best traction, especially on wet roads. A "C" rating will give the least amount of traction. Tread wear ratings use a numbering system ranging from 100 to approximately 500. The higher the number the more mileage can be expected from the tire. A rating of 150 should yield about 50 percent more mileage than a tire rated at 100.

For safety reasons, a periodic tire inspection should be performed. Check for cuts, bulges, abrasions, stone bruises, and objects embedded in the tire. Most tires have tread wear indicators built into the tread. When the wear indicators appear at two adjacent treads at three or more locations around the tire, replace the tire. If the inner cord or fabric is exposed, the tire must be replaced.

3. Measure and adjust tire air pressure.

Overinflation causes wear on the center of the tire tread, and underinflation causes wear on the edges of the tread. Underinflation may also cause tire sidewall damage and wheel damage. Tire pressure should be adjusted when the tires are cool.

Station wagons, trucks, and utility vehicles commonly require more pressure in the rear tires because of the loads they carry and to minimize sway. The pressure should not exceed the maximum specification. Putting more air in the tires on only one side of the vehicle is never recommended.

All air pressure specifications on passenger cars and trucks are to be measured while the tire is cold. A cold tire is one that has been driven fewer than three miles. If the tires have been driven for more than three miles, they must be allowed to cool for two to three hours. Always refer to the vehicle manufacturer's specifications for the tire inflation pressure. Tire pressure specifications can be found on a placard on the driver's door jamb.

4. Diagnose wheel/tire vibration, shimmy, and noise concerns; determine needed repairs.

Tire and wheel vibration problems vary, depending upon the particular problem with the tire and/or wheel. Understanding the characteristics of a road test will help in the diagnosis.

An out of round tire will usually be more noticeable at low speeds and will tend to be less noticeable at highway speeds. A tire which is out of balance will tend to be less noticeable at low speeds and more noticeable at high speeds. A bent wheel or a tire that has lateral runout may cause a shimmy or steering wheel nibble; this may be more noticeable at low speeds. A customer may comment that their hands tingle or go numb after driving their vehicle for awhile when there is a lateral runout concern. A vibration at 55–65 mph, 88 kph–104 kph, is an indication of a tire or wheel imbalance problem. Vibrations which are felt in the seat are usually associated with rear tire/wheel imbalance. Vibrations which are felt in the steering wheel are usually associated with front tire/wheel imbalance. Also, inspect the wheels for damage, cracks, and elongated mounting holes.

Problems within the tire, shifting cords, and damaged tires can cause tires and wheels to become unbalanced. Sometimes a thumping noise can be heard from the tires when the tread becomes damaged or chopped. Wheel tramp is the term used when the tire hops up and down as it rotates. Wheel shimmy refers to the side-to-side shaking that occurs when a tire is not properly balanced. Damaged tires and damaged or bent wheels should be replaced.

5. Rotate tires/wheels and torque fasteners according to manufacturers' recommendations.

Most vehicle manufacturers recommend tire rotation at specified intervals to obtain maximum tire life. The exact tire rotation procedure depends on the model year, the type of tire, and whether the vehicle has a conventional or compact spare. For proper tire rotation procedures, refer to the vehicle manufacturer's service manual or owner's manual. There are several different recommended rotation patterns. The most common method is known as the "modified X." This method involves rotating the drive tires straight forward, or straight rearward as the case may be, and crossing the non-drive tires.

When rotating tires, inspect the sidewalls of the tires for directional indicators; some tires are designed to rotate in a specific direction. Directional tires are typically rotated front to rear to keep the tires rotating in the correct direction. Some vehicles have directional wheels that are designed to work only on one side or one location on the vehicle. Directional tires offer better handling and response when they roll in the intended direction. They are also constructed with a particular tread design that channels away water more effectively, reducing the chances of hydroplaning.

All lug nuts must be tightened to proper torque and in the proper sequence. Consult the service manual for specifications. Overtightening the lug nuts can cause damage to the wheel, distort the wheel studs, or warp the hub/bearing assembly.

6. Measure wheel, tire, axle flange, and hub runout (radial and lateral); determine needed repairs.

Excessive rear chassis waddle may be caused by a shifted steel belt in a tire or a bent rear hub flange. Tire and wheel runout can be checked by using a runout gauge that follows the tire tread (radial runout), or the gauge can be placed on the sidewall of the tire (lateral runout).

If you carefully check runout of all of the parts involved and mark all of the high and low spots, you can correct excessive runout. New parts may not be necessary to correct the problem. Because front-wheel drive cars are lighter and smaller, they transmit noises, vibrations, harshness, and out of round conditions more than larger rear-wheel drive cars and trucks. If overall runout exceeds 0.045–0.060 inch, 1.14mm–1.52mm, there may be a customer concern about vibration or shimmy.

7. Diagnose tire pull (lead) problems; determine corrective actions.

Steering pull may be caused by front tires with different types, sizes, inflation pressure, or tread designs, or a front tire with a conicity defect.

Conicity is a term used in the tire industry when a tire belt is not centered on the tire body. The bead forms a cone, which causes the vehicle to pull in the direction of the small side of the cone. This condition is more commonly referred to by the technician as tire lead. An out of round condition in the rear will cause a vehicle to shake side to side. A front out of round tire will cause the steering wheel to shake, typically at lower speeds.

8. Dismount and mount tire on wheel.

Tires are mounted and dismounted from a wheel using a tire changer. A good tire changer allows the tire to be removed without damaging the tire's beads or the edges of the wheel.

When removing or mounting the tire, apply an approved tire lubricant to the bead area. Otherwise, excessive strain may be put on the tire bead, resulting in damage to the tire.

Clean the sealing area and inspect the wheel for cracks, dents, and burrs. Use a clip-on air chuck while inflating the tire and stand back. Safety glasses should be worn. Do not exceed 40 psi, 276 kPa, in an effort to seat the tire bead. If the bead will not seat, deflate the tire and examine the tire for the cause. After the tire is inflated, adjust to recommended pressure and check for leaks.

9. Balance wheel and tire assembly.

Dynamic wheel balance refers to the balance of a wheel in motion. Cupped tire treads, wheel shimmy, or excessive steering linkage wear may be caused by dynamic wheel imbalance.

A wheel and tire assembly that is statically unbalanced will bounce up and down. A wheel and tire assembly that is dynamically unbalanced will cause the wheel to shake from side to side. This is called a shimmy.

To bring a tire in static balance, weight is added to the rim opposite of the heavy side. The amount of weight necessary is determined by the weight of the heavy section of the tire. The added weight will compensate for the heavy part of the tire and bring the tire into correct static balance.

To be in dynamic balance, the tire must also be in static balance. The tire must be spun to check for dynamic imbalance. The wheel-balancing machine will detect any side-to-side imbalance and determine the location and weight needed to bring the tire in correct dynamic balance. After the weight is added, the tire should be spun again to check for accuracy.

There are many different types of wheel designs used today. Many vehicles are equipped with aluminum wheels. Always use the correct wheel weight designed for the wheel. Using the wrong weight may result in damage to the bead surface and wheel. The wrong design wheel weight may also fail because it may not fit correctly to the contour of the wheel bead area. Many wheel weight manufacturers provide a "rim profile gauge" (and application charts) to help identify the correct wheel weight type to be used.

10. Test and diagnose tire pressure monitoring system (TPMS); determine needed repairs; perform system relearn as required.

The tire pressure monitoring system has a sensor in each tire attached to the inside of the rim to monitor the pressures. These sensors are typically part of the valve stem assembly. The sensors have batteries that are expected to last 10 years. The batteries are not serviceable separate from the sensor, and if they fail, the entire sensor must be replaced.

A light on the dash informs the driver if there is a problem with the tire pressures. When tires are rotated or loss of pressure occurs, the low air pressure light will have to be reset by putting the computer into diagnostic mode according to the manufacturer, and the light reset according to manufacturer's procedures. Most cars have a reset mode that allows you to reset the light by turning the key on and letting a small amount of air out of each tire. There are also aftermarket reset tools that are designed to reset the lights the way a manufacturer would. Extreme care must be taken when mounting/dismounting a tire with a tire pressure monitor. It is easy to damage the sensor. It is recommended that the dismount procedure start away from the sensor. System relearn can be accomplished by using a scan tool, handheld tester, or by simultaneously holding the lock and unlock buttons on the key fob until the horn chirps, indicating the TPMS system is in programming mode.

Sample Preparation Exams

INTRODUCTION

Included in this section are a series of six individual preparation exams that you can use to help determine your overall readiness to successfully pass the Suspension and Steering (A4) ASE certification exam. Located in Section 7 of this book you will find blank answer sheet forms you can use to designate your answers to each of the preparation exams. Using these blank forms will allow you to attempt each of the six individual exams multiple times without risk of viewing your prior responses.

Upon completion of each preparation exam, you can determine your exam score using the answer keys and explanations located in Section 6 of this book. Included in the explanation for each question is the specific task area being assessed by that individual question. This additional reference information may prove useful if you need to refer back to the task list located in Section 4 for additional support.

PREPARATION EXAM 1

1. Which of the following is the proper way to adjust the rack yoke bearing?
   A. Tighten to 75 ft lbs.
   B. Tighten to 5 inch lbs.
   C. Tighten and then back off the original equipment manufacturer's (OEM) specifications.
   D. Tighten to 0.002″ end-play.

2. A customer says that the steering wheel turns more turns to the left than the right. Which of the following is the LEAST LIKELY cause?
   A. Incorrectly timed steering gear
   B. Bent pitman arm
   C. Faulty power steering pump
   D. Bent tie rod

3. A vehicle is hard to steer in one direction only. Which of the following could be the cause?
   A. Tight joint in the steering shaft
   B. Loose joint in the steering shaft
   C. Faulty power steering pump
   D. Faulty power steering gear

4. The rear tires have a diagonal wear pattern. Which of the following is the most likely cause?
   A. Rear toe
   B. Rear camber
   C. Front camber
   D. Front caster

5. A customer says their car makes more noise on asphalt than on concrete road surfaces. Which of the following could be the cause?

    A. Front wheel bearing wear
    B. Rear wheel bearing wear
    C. Aggressive tire tread patterns
    D. Incorrect alignment angles

6. There is a loud roaring noise while driving. The noise gets louder when vehicle speed increases. Which of the following could be the cause?

    A. Worn power steering pump
    B. Worn wheel bearing
    C. Worn ball joint
    D. Worn power steering gear

7. A setback condition is found during an alignment. Which of the following could be the cause?

    A. Worn spring shackle
    B. Worn spring eye bushing
    C. Worn tie rod end
    D. Worn pitman arm

8. There is a whining noise heard only when turning. Which of the following could be the cause?

    A. Leaking internal seals in the steering gear
    B. Worn ball joint
    C. Worn power steering gear
    D. Worn power steering pump

9. The front leaf springs have been replaced on a vehicle. Which of the following operations must also be performed?

    A. Vehicle alignment
    B. Tire rotation
    C. Tire balance
    D. Rear spring replacement

10. A vehicle has worn jounce bumpers. Which of the following could be the cause?

    A. Leaking shocks
    B. Leaking power steering pump
    C. Leaking power steering gear
    D. Worn strut bearings

11. A worn idler arm would affect which alignment angle?

    A. Caster
    B. Camber
    C. Toe
    D. SAI

12. A shifted front cradle could affect any of these EXCEPT:

    A. Front camber.
    B. Front caster.
    C. Thrust line.
    D. Thrust angle.

13. There is a pull on acceleration and deceleration. Technician A says spring eye bushings could be the cause. Technician B says worn strut rod bushings could be the cause. Who is right?

    A. A only
    B. B only
    C. Both A and B
    D. Neither A nor B

14. A composite leaf spring is being installed in a vehicle. Which of the following is correct?

    A. The spring can be installed using a pry bar.
    B. The spring can be compressed with vise grips.
    C. The spring should be painted after installation.
    D. The spring should not be scratched during installation.

15. Which of the following alignment specifications would most likely need a steering damper?

    A. 6 degrees caster
    B. 2 degrees camber
    C. 2 degrees caster
    D. 6 degrees camber

16. A customer complains of a popping noise while turning the steering wheel. Which of the following is the most likely cause?

    A. Rear shock mounts
    B. Rear toe adjusters
    C. Front shock mounts
    D. Front tie rods

17. A power steering pump hose is leaking on a vehicle. Which of the following is the most likely cause?

    A. Separated motor mounts
    B. Loose drive belt
    C. Air in the fluid
    D. Broken pump mounts

18. Which angle has the greatest affect on steering wheel return?

    A. Camber
    B. SAI
    C. Toe
    D. Caster

19. There is a vibration which can be felt in the steering wheel at speeds above 30 mph. Which of the following could be the cause?

    A. Front wheels out of balance
    B. Rear wheels out of balance
    C. Caster
    D. Camber

20. The hole in the steering knuckle into which the ball joint stud goes is out of round. Which of the following is the best method of repair?

    A. Measure the wear and check against manufacturer's specifications.

    B. Replace the steering knuckle.

    C. Ream the hole.

    D. Shim the hole.

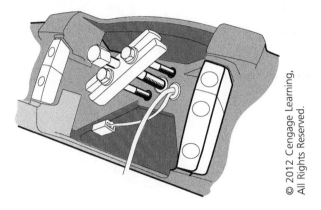

© 2012 Cengage Learning, All Rights Reserved.

21. The setup shown in the photo above is used to:

    A. Remove the horn.

    B. Remove the steering wheel.

    C. Install the air bag.

    D. Install the steering wheel.

22. The left front tire on a vehicle shows outside tread wear. Which of the following is the most likely cause?

    A. Worn shock absorbers

    B. Tires overinflated

    C. Camber adjustment incorrect

    D. Caster adjustment incorrect

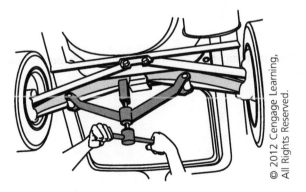

© 2012 Cengage Learning, All Rights Reserved.

23. What is being performed in the photo above?

    A. Installing a longitudinally mounted leaf spring

    B. Removing a transversely mounted leaf spring

    C. Installing a coil spring

    D. Removing a coil spring

24. The tools shown in the photo above are used to:

    A. Remove seals.

    B. Install seals.

    C. Remove bearings.

    D. Install bearings.

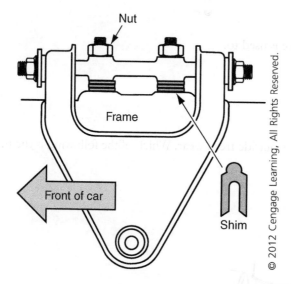

25. Referring to the figure above, if shims are removed from the rear shim pack and installed in the front spring pack, which of the following will occur?

    A. Toe will go positive.

    B. Caster will go negative.

    C. Caster will go positive.

    D. Toe will go negative.

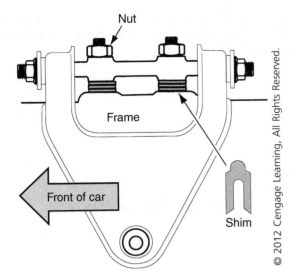

26. Referring to the figure above, if shims are added to both shim packs, which of the following will occur?

    A. Toe will go positive.
    B. Camber will go negative.
    C. Camber will go positive.
    D. Toe will go negative.

27. Which of the following would cause toe-out-on-turns to be out of specification?

    A. Bent tie rod end
    B. Bent steering arm
    C. Bent lower control arm
    D. Bent upper control arm

28. A vehicle is dog-tracking. Which alignment angle is most likely the cause?

    A. Rear toe
    B. Front toe
    C. Rear camber
    D. Front camber

29. A vehicle is equipped with an air ride suspension. The vehicle will not lower when necessary. Which of the following could cause this problem?

    A. Compressor relay
    B. Compressor
    C. Sensor
    D. Fuse

30. Which of the following is the correct way to adjust tapered roller bearings?

    A. Torque and then back off OEM specification.
    B. Torque to OEM specification.
    C. Adjust to a minimum of .150″ (3.80 mm) end-play.
    D. Torque to 50 ft lbs.

31. A vehicle has a crooked steering wheel when driving down the road. Technician A says this could be caused by failure to install the steering wheel correctly. Technician B says this could be caused by rear toe being out of adjustment. Who is correct?

    A. A only
    B. B only
    C. Both A and B
    D. Neither A nor B

32. When does a positive camber angle exist?

    A. When the top of the tire leans outward
    B. When the top of the tire leans inward
    C. When the upper ball joint is rearward of the lower ball joint as viewed from the side
    D. When the rear axle is crooked

33. Why is the camber angle important?

    A. Because it directly affects ride stiffness
    B. Because it is a tire wearing angle and it affects steering control
    C. Because it can cause vibration
    D. Because it affects the operation of the brakes

34. The caster angle on the left front wheel has been set at 1 degree negative. The right front caster angle has been set at 1 degree positive. What would be the most likely result?

    A. The car would pull to the right.
    B. The car would pull to the left.
    C. The car would travel straight ahead.
    D. The left tire would show second rib wear.

35. Technician A says that camber and SAI make up the included angle. Technician B says that the purpose of the turning angle is to reduce tire scuffing during cornering. Who is right?

    A. A only
    B. B only
    C. Both A and B
    D. Neither A nor B

36. Technician A says that shims are sometimes used to adjust rear toe and camber. Technician B says that strut rods are sometimes used to adjust rear toe and camber. Who is right?

    A. A only
    B. B only
    C. Both A and B
    D. Neither A nor B

37. When the rear of the rear tires are closer together than at the front, the tires have:

    A. Negative camber.
    B. Positive camber.
    C. Toe-in.
    D. Toe-out.

38. The jounce bumpers on the rear of a vehicle are severely worn. Which of the following is the LEAST LIKELY cause?

    A. Worn rear springs
    B. Excessive cargo in the trunk
    C. Worn front springs
    D. Worn rear shocks

39. A customer has a noise concern coming from the rear of the vehicle. Any of these could be the cause EXCEPT:

    A. Worn rear wheel bearings.
    B. Rear tires out of balance.
    C. Missing spring separators.
    D. Worn spring eye bushings.

40. Which of the following is true concerning torsion bar front suspension?

    A. Torsion bars must always be replaced in pairs.
    B. Torsion bars are non-adjustable.
    C. When torsion bars are reinstalled, they can be installed in either side.
    D. Torsion bars can be used on SLA suspensions.

PREPARATION EXAM 2

1. Which of the following could cause poor returnability?

    A. Thrust angle
    B. Toe-out-on-turns
    C. Tire pressure
    D. Power steering pump pressure

2. Which of the following alignment angles is most likely to be the cause of a steering returnability concern?

    A. Front toe
    B. Caster
    C. Camber
    D. Rear toe

3. Technician A says the SAI/KPI (steering axis inclination/key performance indicator) angle can be used to diagnose a bent strut. Technician B says the SAI/KPI angle can be used to diagnose a bent steering arm. Who is correct?

    A. A only
    B. B only
    C. Both A and B
    D. Neither A nor B

4. Any of these would be an indication of a front cradle out of alignment EXCEPT:

    A. SAI/KPI varies 2 degrees side to side.
    B. Camber positive on the left side and negative on the right side.
    C. Camber negative on the left side and positive on the right side.
    D. Thrust angle of 0.1 degrees.

5. The owner of a vehicle equipped with manual steering complains of hard steering. The alignment is checked and the readings are below:

	Spec	Actual	
		Left	Right
Caster	0.0 degrees	2.0 degrees	2.1 degrees
Camber	0.5 degrees	0.6 degrees	0.5 degrees
Toe	0.5 degrees	0.2 degrees	0.2 degrees

Which of the following is the most likely cause?

A. Camber is too negative.
B. Caster is too negative.
C. Camber is too positive.
D. Caster is too positive.

6. Refer to the alignment readings below:

	Spec	Actual		Tolerance
		Left	Right	
Caster	0.0 degrees	0.0 degrees	0.1 degrees	+/− 0.2
Camber	0.5 degrees	0.6 degrees	0.5 degrees	+/− 0.2
Toe	0.4 degrees	0.2 degrees	0.2 degrees	+/− 0.1

Which of the following is a true statement?

A. Caster should be adjusted.
B. Camber should be adjusted.
C. Caster and camber are within tolerance and toe should be adjusted.
D. This vehicle is within alignment tolerances.

7. A vehicle with the following alignment readings is wearing the outside edge of the right front tire. Which of the following is most likely the cause?

	Spec	Actual		Tolerance
		Left	Right	
Caster	0.0 degrees	2.0 degrees	0.1 degrees	+/− 0.2
Camber	0.5 degrees	0.6 degrees	3.5 degrees	+/− 0.2
Toe	0.5 degrees	0.3 degrees	0.2 degrees	+/− 0.1

A. Left toe
B. Left caster
C. Right camber
D. Right toe

8. A four-wheel drive vehicle with oversized tires and rims has a shimmy after hitting a bump. A dry park inspection has been performed and the alignment has been checked. No problems were identified. Which of the following would most likely fix the concern?

A. Installation of a steering damper
B. Adjust the caster 2 degrees positive from the original caster specifications.
C. Adjust the caster 2 degrees negative from the original caster specifications.
D. Installation of a power steering cooler

Delmar, Cengage Learning ASE Test Preparation

9. A vehicle has a shimmy when the steering wheel returns to center after completing a turn. Which of the following could be the cause?

    A. Excessive positive camber
    B. Excessive positive caster
    C. Excessive toe-out
    D. Excessive toe-in

10. Which alignment angle is usually considered the greatest tire wearing angle?

    A. Caster
    B. Camber
    C. Toe
    D. SAI/KPI

11. Which of the following would be the LEAST LIKELY cause of worn jounce bumpers?

    A. Worn shocks
    B. Worn springs
    C. Worn ball joints
    D. A low ride height setting

12. A new composite leaf spring is being installed. Which of the following is true?

    A. The spring must be coated with a penetrant prior to installation.
    B. The spring must be steam cleaned prior to installation.
    C. A ball joint separator must be used to slide the spring in to position.
    D. A spring compressor must be used to install the spring.

13. The power steering pressure switch is used to:

    A. Increase power steering pressure while driving in a straight line.
    B. Prevent stalling while idling at a stop.
    C. Prevent stalling during a parking maneuver.
    D. Increase power steering pressure while turning.

14. A vehicle pulls to the right only while accelerating. Which of the following could be the cause?

    A. Underinflated tires
    B. Loose cradle
    C. Loose power steering pump belt
    D. Overinflated tires

15. A vehicle pulls to the right only while braking. Which of the following could be the cause?

    A. Underinflated tires
    B. Loose cradle
    C. Loose power steering pump belt
    D. Overinflated tires

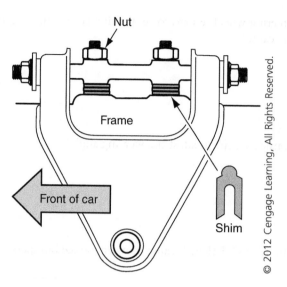

16. Referring to the figure above, what would be the proper method to add caster without changing camber?

    A. Add shims to both sides.
    B. Remove shims from both sides.
    C. Remove shims from the rear shim pack and add an equal thickness of shims to the front shim pack.
    D. Remove shims from the front shim pack and add an equal thickness of shims to the rear shim pack.

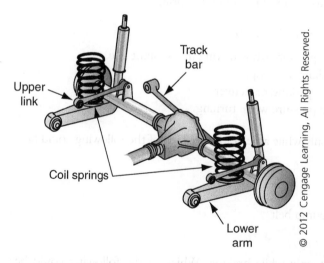

17. A vehicle equipped with the suspension shown in the figure above pulls to the right on acceleration. Technician A says this could be caused by worn track bar bushings. Technician B says this could be caused by worn bushings in the lower arms. Who is correct?

    A. A only
    B. B only
    C. Both A and B
    D. Neither A nor B

18. There is not enough adjustment left to correct rear camber and rear toe on a vehicle. Which of the following could be the cause?

    A. Sagging rear springs
    B. Worn shock absorbers on the front
    C. Worn shock absorbers on the rear
    D. Sagging front springs

19. A vehicle has hard steering at low speeds, and occasionally the engine will stall. Which of the following is the most likely cause?

    A. Faulty power steering pressure switch
    B. Weak power steering pump
    C. Worn steering gear
    D. Worn ball joints

20. Which of the following would most likely make a vehicle hard to steer?

    A. Overtightened power steering pump belt
    B. Crimped power steering line
    C. Positive camber on the rear
    D. Negative caster on the front

21. A vehicle is hard to steer to the left only. Which of the following is the most likely cause?

    A. Weak power steering pump
    B. Tight steering shaft universal joint
    C. Binding ball joint
    D. Worn power steering gear

22. A vehicle is hard to steer. After disconnecting both outer tie rods, the steering wheel is still hard to turn. Which of the following could be the cause?

    A. Binding strut bearing on the left strut
    B. Worn strut bearing on the right strut
    C. Binding steering column universal joint
    D. Worn rack bushings

23. There is a loud growling noise only when the vehicle is steered to the right while driving. Which of the following could be the cause?

    A. Worn tires
    B. Worn wheel bearing
    C. Incorrect camber
    D. Incorrect toe

24. A vehicle has excessive (6 degrees) negative camber on the left front. The right front camber is within specification. Which of the following could be the cause?

    A. A worn right rear spring
    B. A worn left rear spring
    C. A worn right front wheel bearing
    D. A worn left front wheel bearing

25. A vehicle has 1.5 degrees of cross caster. Technician A says this could cause tire wear. Technician B says this could cause a pull. Who is correct?

    A. A only
    B. B only
    C. Both A and B
    D. Neither A nor B

26. A vehicle has 2.5 degrees cross camber. Technician A says this could cause tire wear. Technician B says this could cause a pull. Who is correct?

    A. A only
    B. B only
    C. Both A and B
    D. Neither A nor B

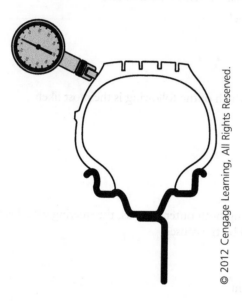

27. What check is being performed in the figure above?
    A. Lateral runout check
    B. Radial runout check
    C. Static balance check
    D. Dynamic balance check

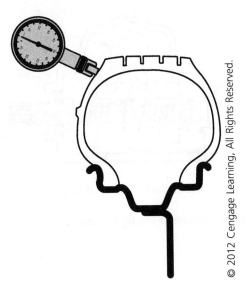

28. Which of the following would be a normal maximum specification for the measurement shown in the figure above?

    A. 0.000″ (0.00 mm)
    B. 0.045″ (1.143 mm)
    C. 0.090″ (2.28 mm)
    D. 0.135″ (3.43 mm)

29. The drive axle nut on a front-wheel drive vehicle is being tightened. Which of the following is true?

    A. The vehicle should be on a frame hoist.
    B. The nut should be tightened, then backed off 1 flat.
    C. The nut should be tightened with an impact wrench.
    D. The vehicle should be at normal ride height.

30. Which steering gear box adjustment should be performed first?

    A. Sector shaft preload
    B. Worm bearing preload
    C. Sector shaft over center
    D. Worm bearing end-play

31. A vehicle sets low on one front corner. Which of the following is the most likely cause?

    A. Weak spring
    B. Weak shock
    C. Bent control arm
    D. Bent steering knuckle

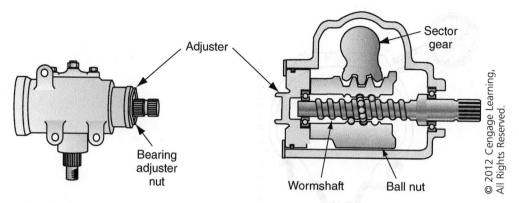

32. Which adjustment is performed with the adjuster shown in the figure above?

    A. Sector shaft preload

    B. Over center adjustment

    C. Worm bearing preload

    D. Sector shaft end-play

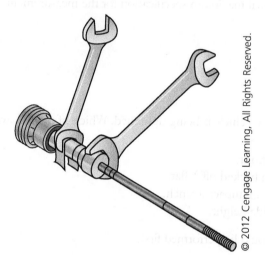

33. What item is being removed in the figure above?

    A. Outer tie rod

    B. Inner tie rod

    C. Pitman arm

    D. Idler arm

34. A customer complains of a vibration only while braking. Technician A says the wheel lug nuts may have been overtorqued. Technician B says the front wheel may be bent. Who is correct?

    A. A only

    B. B only

    C. Both A and B

    D. Neither A nor B

35. Which of the following is the most likely cause of tires which are worn in the center of the tread?

    A. Underinflation

    B. Overinflation

    C. Positive caster

    D. Positive camber

36. Technician A says an ohmmeter can be used to measure the resistance of the airbag inflator module. Technician B says a test light should be used to check for voltage at the airbag inflator module. Who is correct?

    A. A only
    B. B only
    C. Both A and B
    D. Neither A nor B

37. A lower ball joint is being inspected on a front suspension. The front coil spring is located between the frame and the lower control arm. Where must the jack stand be placed?

    A. Under the lower control arm
    B. Under the upper control arm
    C. Under the frame
    D. Under the rear cradle

38. A rear tie rod end is being replaced. Which of the following alignment angles will most likely need to be adjusted?

    A. Front toe
    B. Rear toe
    C. Front camber
    D. Front caster

39. A rear-wheel drive solid rear axle vehicle has a sheared spring center bolt on the left side. Which of the following will need to be replaced?

    A. The entire left rear spring pack
    B. Both the left and right rear spring packs
    C. The left center bolt
    D. Both left and right center bolts

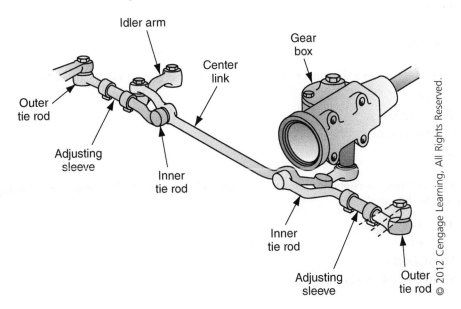

40. The steering linkage illustrated in the figure above is called:

    A. Rack and pinion.
    B. Parallelogram.
    C. Haltenberger.
    D. Relay rod.

PREPARATION EXAM 3

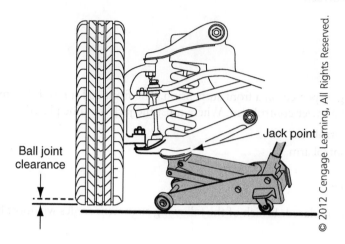

Ball joint clearance

Jack point

1. The figure above is measuring:

    A. The follower ball joint for wear.
    B. The load carrying ball joint for wear.
    C. The front coil spring spacer for wear.
    D. The lower control arm bushings for wear.

2. The left front tire is worn on the inside tread only; the right front tire shows no abnormal wear. Which angle is most likely the cause?

    A. Caster
    B. Camber
    C. Toe-in
    D. Toe-out

3. The right rear tire has a diagonal swipe wear pattern. This is most likely caused by:

    A. Caster.
    B. Positive camber.
    C. Negative camber.
    D. Toe.

4. The left front tire has cupping on both the inner and outer treads. Which of the following is the most likely cause?

    A. Toe-in
    B. Toe-out
    C. Worn shocks
    D. Worn control arm bushings

5. Thrust line is adjusted by changing:

    A. Rear camber.
    B. Front camber.
    C. Front toe.
    D. Rear toe.

6. Technician A says rear camber may be adjusted using shims. Technician B says rear toe may be adjusted using shims. Who is correct?

    A. A only
    B. B only
    C. Both A and B
    D. Neither A nor B

7. Thrust angle is calculated using:

    A. Thrust line and geometric center line.
    B. Camber and SAI.
    C. Thrust line and SAI.
    D. Geometric center line and camber.

8. Any of these can cause bump steer EXCEPT:

    A. Worn idler arm.
    B. Unlevel steering rack.
    C. Incorrect camber settings.
    D. Worn center link.

9. The tire pressure monitor light on the dash is illuminated. Which of the following is the LEAST LIKELY cause?

    A. Excessive tire pressure in the right rear tire
    B. Low tire pressure in the left front tire
    C. Low tire pressure in the spare
    D. A faulty tire pressure sensor

10. Which of the following is true about tire pressure sensors?

    A. The batteries are designed to last a minimum of 15 years.
    B. The batteries are separate replaceable units.
    C. The sensor should be replaced every time the tires are replaced.
    D. The sensors can be part of the valve stem assembly.

11. A customer has a concern about low power assist during parallel park maneuvers. The technician verifies the concern and notes a growling noise coming from the pump. Which of the following is the most likely cause of the concern?

    A. Loose power steering belt
    B. Low power steering fluid
    C. Internally restricted power steering fluid cooler
    D. Externally restricted power steering fluid cooler

12. A suspension bushing was tightened while the vehicle was supported by a frame hoist. This will cause:

    A. Camber wear on the tires.
    B. Toe wear on the tires.
    C. Short bushing life.
    D. The vehicle to pull while driving straight.

13. A front-wheel drive vehicle makes a loud clunk on initial acceleration. Any of these could be the cause EXCEPT:

    A. Worn inner CV joint.
    B. Worn rack mount bushings.
    C. Worn control arm bushings.
    D. Loose cradle mounting bolts.

14. The technician finds the steering wheel moves side to side within the steering column. Which of the following could be the cause?

    A. Worn rack mounts
    B. Worn tie rod ends
    C. Worn upper steering column bearing
    D. Loose steering column mounts

15. A vehicle equipped with electric power steering (EPS) has been aligned. Now the vehicle pulls to the left while traveling on a straight, level road. Technician A says this could be caused by an incorrect front toe setting. Technician B says the EPS may need to be re-centered. Who is correct?

    A. A only
    B. B only
    C. Both A and B
    D. Neither A nor B

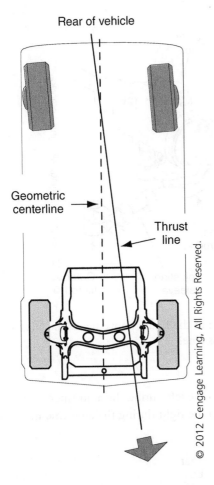

Rear of vehicle

Geometric centerline →

Thrust line

16. What adjustment must be performed to correct the alignment condition shown in the figure above?

    A. Rear toe

    B. Front toe

    C. Rear camber

    D. Front camber

17. While making a hard brake application, the customer notices a clunk from the rear of the vehicle. Which of the following could be the cause?

    A. Leaking struts

    B. A bent rim

    C. An out of round tire

    D. Worn lower control arm bushings

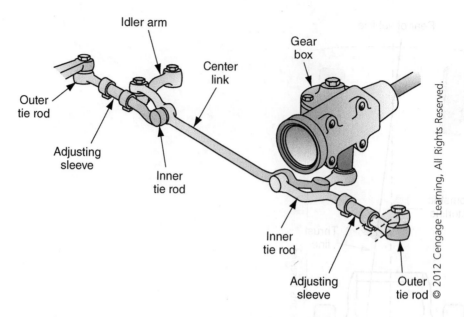

18. A vehicle with the steering/linkage system shown in the figure above needs the left toe adjusted to the positive. Which of the following is true?

   A. The adjusting sleeve must be lengthened.
   B. The adjusting sleeve must be shortened.
   C. The steering should be turned 20 degrees to the left during the adjustment.
   D. The steering should be turned 20 degrees to the right during the adjustment.

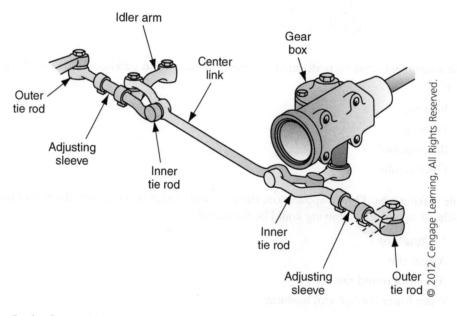

19. In the figure above, the technician is preparing to turn the adjusting sleeve. Which of the following statements is true?

   A. This should be done prior to rear toe.
   B. This should be done prior to rear camber.
   C. This should be done prior to front camber.
   D. This should be done as the last adjustment.

20. A vehicle with the alignment angles shown below is pulling to the left. Which of the following is the most likely cause?

	Spec	Actual		Tolerance
		Left	**Right**	
Caster	3.9 degrees	3.8 degrees	4.8 degrees	+/− 0.8
Camber	−0.4 degrees	−0.2 degrees	−0.6 degrees	+/− 0.8
Toe	0.2 degrees	0.15 degrees	−0.05 degrees	+/− 0.2

A. Right caster
B. Left caster
C. Left camber
D. Right camber

21. The vehicle with the alignment readings shown below has a crooked steering wheel. Which of the following is the most likely cause?

FRONT

	Spec	Actual		Tolerance
		Left	**Right**	
Caster	1.4 degrees	1.5 degrees	1.3 degrees	+/− 0.2
Camber	0.5 degrees	0.6 degrees	0.4 degrees	+/− 0.2
Toe	0.5 degrees	0.4 degrees	0.4 degrees	+/− 0.1

REAR

	Spec	Actual		Tolerance
		Left	**Right**	
Camber	0.5 degrees	0.6 degrees	0.3 degrees	+/− 0.2
Toe	0.0 degrees	−0.3 degrees	0.2 degrees	+/− 0.1

A. Rear camber
B. Front camber
C. Front toe
D. Rear toe

22. A vehicle has hard steering at low speeds, and occasionally the engine will stall while parallel parking. Which of the following is the most likely cause?

A. Dirty throttle body
B. Weak power steering pump
C. Worn steering gear
D. Worn ball joints

23. Replacing which of the following will change ride height?

A. Shocks
B. Struts
C. Springs
D. Sway bars

24. A steering wheel must be removed on a vehicle equipped with a supplemental restraint system (SRS). Technician A says the SRS must have the power removed and the appropriate wait period as prescribed in the service manual must be observed. Technician B says SRS airbag removal can begin as soon as the SRS fuse is removed. Who is correct?

    A.   A only

    B.   B only

    C.   Both A and B

    D.   Neither A nor B

25. When storing an airbag on a workbench, which of the following is the correct way to place it?

    A.   Face down

    B.   Face up

    C.   Facing the wall

    D.   Facing away from the wall

26. The power steering pressure is being measured. The fluid is warm, the engine is idling, the tester valve is open, and the steering wheel is in the straight ahead position. Which of the following would be considered an acceptable reading?

    A.   0 psi

    B.   100 psi

    C.   500 psi

    D.   1,000 psi

27. The center link is unlevel on a vehicle with parallelogram steering. Which of the following could be the cause?

    A.   Incorrect idler arm height adjustment

    B.   Incorrect worm bearing adjust

    C.   Incorrect sector shaft adjustment

    D.   Incorrect caster adjustment

28. Which of the following would be used to remove a pitman arm?

    A.   Pickle fork

    B.   Ball joint separator

    C.   Puller

    D.   Slide hammer

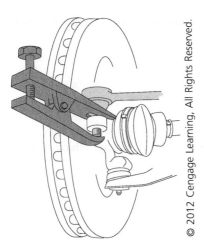

29. What is being performed in the figure above?

    A. Idler arm removal

    B. Pitman arm removal

    C. Ball joint removal

    D. Tie rod end removal

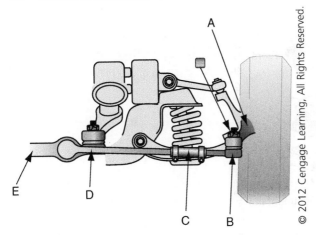

30. Item D in the figure above was bent during an accident. Which of the following would occur?

    A. Toe would go positive.

    B. Toe would go negative.

    C. Caster would go positive.

    D. Caster would go negative.

31. Both front tires on a solid front axle vehicle have excessive positive camber. Which of the following could be the cause?

    A. Caster shims were installed.
    B. The toe was improperly adjusted.
    C. The axle is bent.
    D. The drag link is bent.

32. The left front leaf spring is being replaced on a solid front axle vehicle. Which of the following should also be changed?

    A. The left front shock
    B. Both front shocks
    C. The right front leaf spring
    D. The left rear spring

33. A customer is concerned that the rear suspension sags when there are passengers in the rear seats. Which of the following is the most likely cause?

    A. Weak shocks
    B. Worn sway bar
    C. Worn jounce bumpers
    D. Weak springs

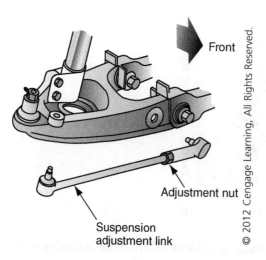

34. The adjustment link shown in the figure above is worn. This would most likely result in:

    A. A change in front toe.
    B. A change in front camber.
    C. A change in rear toe.
    D. A change in rear camber.

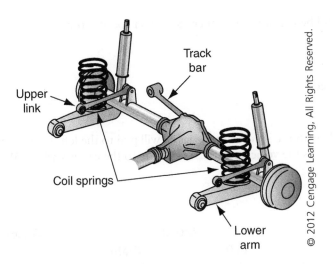

Track bar

Upper link

Coil springs

Lower arm

35. A vehicle with the suspension system shown in the figure above has excessive driveline vibrations only when loaded. Which of the following could be the cause?

   A.   Worn track bar bushings
   B.   A bent track bar
   C.   A bent lower control arm
   D.   Worn upper link bushings

36. A solid rear axle vehicle has a shorter wheelbase on one side than the other. Technician A says this could cause a pull to the side with the shorter wheelbase. Technician B says this could cause a thrust angle problem. Who is correct?

   A.   A only
   B.   B only
   C.   Both A and B
   D.   Neither A nor B

37. Which of the following would be the most normal end-play specification for a set of tapered roller front-wheel bearings?

   A.   0.0001″–0.0005″ (0.00254 mm–0.0127 mm)
   B.   0.0010″–0.0050″ (0.0254 mm–0.127 mm)
   C.   0.0100″–0.0500″ (0.254 mm–1.27 mm)
   D.   0.1000″–0.5000″ (2.54 mm–12.7 mm)

38. Which of the following would be the most normal cross caster specification?

   A.   No more than 0.5 degrees
   B.   No less than 0.5 degrees
   C.   No more than 1.5 degrees
   D.   No less than 1.5 degrees

39. Which of the following would be the most normal cross camber specification?

    A. No more than 0.5 degrees
    B. No less than 0.5 degrees
    C. No more than 1.5 degrees
    D. No less than 1.5 degrees

40. Technician A says the vehicle with the specifications below will pull to the left due to the caster settings. Technician B says the vehicle will wear tires due to the caster settings. Who is correct?

FRONT

	Spec	Actual		Tolerance
		Left	Right	
Caster	2.4 degrees	1.5 degrees	1.3 degrees	+/– 0.5
Camber	0.5 degrees	0.5 degrees	0.4 degrees	+/– 0.2
Toe	0.5 degrees	0.25 degrees	0.25 degrees	+/– 0.2

REAR

	Spec	Actual		Tolerance
		Left	Right	
Camber	0.5 degrees	0.6 degrees	0.5 degrees	+/– 0.2
Toe	0.0 degrees	–0.1 degrees	0.1 degrees	+/– 0.1

    A. A only
    B. B only
    C. Both A and B
    D. Neither A nor B

PREPARATION EXAM 4

1. The tire pressure monitor system (TPMS) light is illuminated on the dash. Which of the following could be the cause?

    A. The traction control system has a code.
    B. One tire has low tire pressure.
    C. The torsion bars sense an overload.
    D. The tires are due to be rotated.

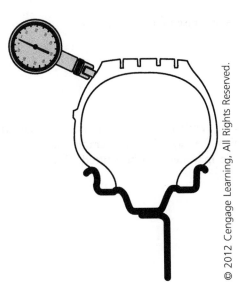

2. The measurement shown in the figure above is out of specification. Technician A says an out of round brake drum could be the cause. Technician B says a bent rim could be the cause. Who is correct?

   A. A only
   B. B only
   C. Both A and B
   D. Neither A nor B

3. An out of round rear tire would cause what type of customer concern?

   A. A vibration that tends to gets worse at higher road speeds
   B. A vibration in the steering wheel
   C. A vibration which is worse at low speeds and tends to get better at high speeds
   D. A vibration that only occurs during cornering

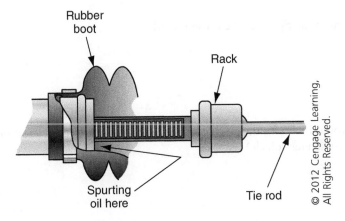

4. Fluid is spurting as shown in the figure above. Which of the following is the most likely repair procedure?

   A. Replace the tie rod.
   B. Replace the rubber boot.
   C. Replace the inner rack seal.
   D. Replace the rack.

5. Which of the following would be the most likely cause of a steering pull?

    A. Worn control arm bushings

    B. Broken sway bar end links

    C. Worn shock absorber bushings

    D. Incorrect front toe settings

6. A vehicle alignment is checked. The left rear camber is excessively positive. The right rear camber is excessively negative. Technician A says the front cradle may need to be shifted. Technician B says the rear cradle may need to be shifted. Who is correct?

    A. A only

    B. B only

    C. Both A and B

    D. Neither A nor B

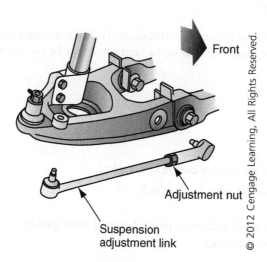

Front

Adjustment nut

Suspension adjustment link

7. The adjustment link shown in the figure above is used to adjust:

    A. Front camber.

    B. Front toe.

    C. Rear toe.

    D. Rear camber.

8. Which of the following would be the most likely maximum radial runout specification on a wheel tire assembly for a passenger vehicle?

    A. 0.020″

    B. 0.030″

    C. 0.050″

    D. 0.090″

9. A technician is preparing to measure radial runout on a wheel tire assembly. Where should the dial indicator be placed?

   A. On the rim bead seat parallel to the axle
   B. On the rim bead seat perpendicular to the axle
   C. On the tire parallel to the axle
   D. On the tire perpendicular to the axle

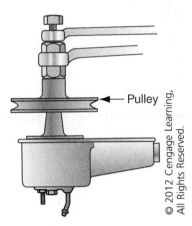

Pulley

10. Which procedure is being performed in the figure above?

    A. Replacing a power steering pump pulley
    B. Adjusting power steering pump belt tension
    C. Mounting the power steering pump
    D. Preparing the pump for pressure testing

11. Tapered roller wheel bearings are being serviced. Which of the following is true?

    A. The cup is always replaced.
    B. The hub must be held still while the bearing is adjusted.
    C. The seal is always replaced.
    D. The bearing should be adjusted with the tire on the ground.

12. A tie rod end has been installed and the nut properly torqued. The cotter pin hole is not aligned. Which of the following should the technician do?

    A. Replace the tie rod end.
    B. Replace the steering knuckle.
    C. Tighten the nut until the hole aligns.
    D. Loosen the nut until the hole aligns.

13. There is excessive play in the steering system. Which of the following is the most likely cause?

    A. Worn strut mounts
    B. Worn sway bar bushings
    C. Worn shock mounts
    D. Worn rack mounts

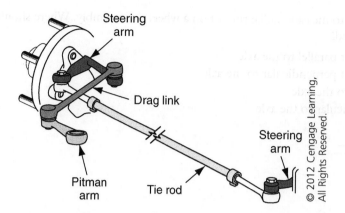

14. Referring to the figure above, which of the following would be used to adjust total toe?

    A.   Tie rod
    B.   Steering arm
    C.   Drag link
    D.   Pitman arm

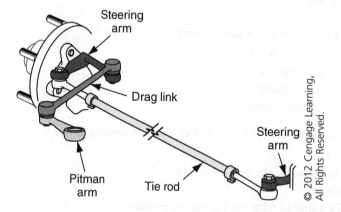

15. Referring to the figure above, which of the following would be used to center the steering wheel?

    A.   Tie rod
    B.   Steering arm
    C.   Drag link
    D.   Pitman arm

16. The steering binds while turning in either direction. The pitman arm is removed and the steering continues to bind. Which of the following could be the cause?

    A.   A binding steering shaft U-joint
    B.   A binding upper ball joint
    C.   A binding lower ball joint
    D.   A binding tie rod end

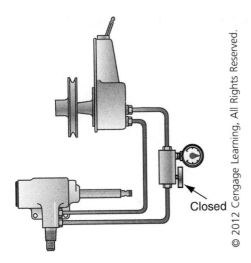

Closed

17. What is being measured when the valve illustrated in the figure above is closed?

   A. Fluid flow (GPH)
   B. Pump vacuum (Hg)
   C. Pump pressure (PSI)
   D. Fluid temperature (F)

18. The steering knuckle is being replaced on the front of a vehicle with torsion bar SLA suspension. Any of these is true concerning this procedure EXCEPT:

   A. The outer tie rod must be disconnected.
   B. The inner tie rod must be disconnected.
   C. The upper ball joint must be disconnected.
   D. The lower ball joint must be disconnected.

19. Any of these is true about ball joint replacement EXCEPT:

   A. The ball joint may be pressed in.
   B. The ball joint may be bolted in.
   C. The control arm may have to be replaced with the ball joint.
   D. The inner tie rod may have to be replaced with the ball joint.

20. The sector shaft adjustment is being performed on a recirculating ball steering gear. Technician A says turning torque should be measured with an inch/pound torque wrench. Technician B says the adjustment is changed by rotating the adjusting screw on the top of the steering gear. Who is correct?

   A. A only
   B. B only
   C. Both A and B
   D. Neither A nor B

21. An idler arm is being replaced. Which of the following alignment angles will need to be checked after replacement?

    A. Caster
    B. Toe
    C. Camber
    D. Thrust line

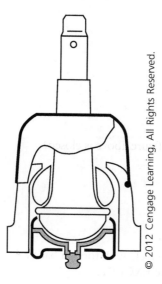

22. The ball joint shown in the figure above is being inspected. Technician A says the vehicle will need to be lifted on a frame hoist. Technician B says the ball joint must be removed from the vehicle to be properly inspected. Who is correct?

    A. A only
    B. B only
    C. Both A and B
    D. Neither A nor B

23. A vehicle becomes unpredictable when cornering. Which of the following is the most likely cause?

    A. Worn track bar bushings
    B. Worn shock bushings
    C. Weak front springs
    D. Weak rear springs

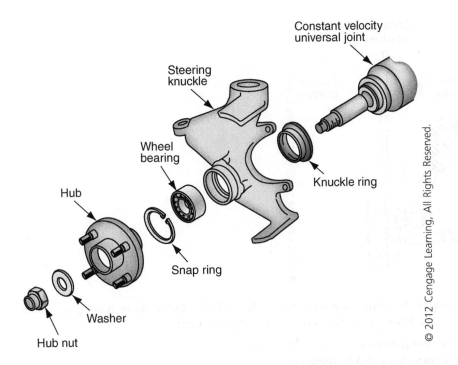

24. The bearing illustrated in the figure above is being reinstalled. Which of the following is correct concerning installing the hub nut?

    A. The hub nut can be reused.

    B. The hub nut should be torqued with the wheels on the ground.

    C. The hub nut should be torqued, then backed off one-half turn.

    D. Locking compound should be used on the hub nut.

25. Technician A says ride height must be measured both before and after an alignment. Technician B says ride height should be measured before an alignment. Who is correct?

    A. A only

    B. B only

    C. Both A and B

    D. Neither A nor B

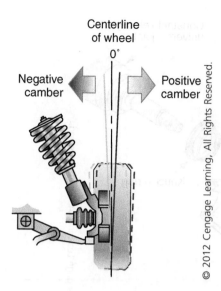

26. The spring in the front suspension system shown in the figure above is being replaced. Which of the following is true concerning this procedure?

    A. The spring is compressed on the vehicle.
    B. The strut must also be replaced.
    C. The alignment will need to be checked after spring replacement.
    D. The rear spring on the same side must also be replaced.

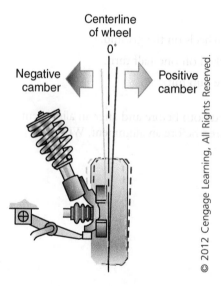

27. The spring in the front suspension system shown in the figure above is being replaced. Which of the following is true concerning this procedure?

    A. The spring is compressed on the vehicle.
    B. The strut must also be replaced.
    C. The rear alignment toe angle will be changed.
    D. The front spring on the other side must also be replaced.

28. A vehicle with coil spring front suspension has a low ride height. Technician A says the springs can be adjusted to correct ride height. Technician B says the springs can be rotated to correct ride height. Who is correct?

    A. A only
    B. B only
    C. Both A and B
    D. Neither A nor B

29. A customer has requested that the rear struts be replaced on their vehicle. The vehicle has McPherson strut independent rear suspension. Which of the following is true?

    A. The strut bearings must also be replaced.
    B. The rear springs must also be replaced.
    C. The alignment must be corrected prior to replacement.
    D. The rear alignment must be checked after installation.

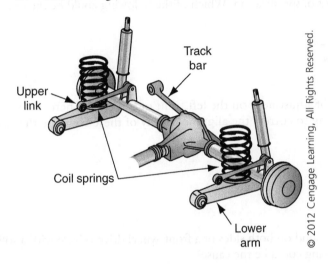

30. The shocks are being replaced on the suspension system shown in the figure above. Which of the following is true regarding this procedure?

    A. The upper links will need to be disconnected.
    B. The lower arms will need to be disconnected.
    C. If a frame hoist is used, the axle will need to be supported.
    D. The track bar must also be replaced.

31. A jounce bumper on the rear suspension is worn excessively. Which of the following is the most likely cause?

    A. Incorrect toe settings
    B. Incorrect camber settings
    C. Worn springs
    D. Worn strut rod bushings

32. An alignment is being performed on a McPherson strut vehicle. Technician A says the toe adjustment may be at the base of the strut. Technician B says the camber adjustment may be at the strut mount. Who is correct?

    A. A only
    B. B only
    C. Both A and B
    D. Neither A nor B

33. A vehicle has a negative setback. Which of the following is true?

    A. The vehicle may pull to the left.
    B. The vehicle may pull to the right.
    C. The left rear tire will have a diagonal swipe wear pattern.
    D. The right rear tire will have an outside tread wear pattern.

34. Steering axis inclination is out of specification. Which of the following could be the cause?

    A. A bent steering arm
    B. Incorrect camber settings
    C. Incorrect caster settings
    D. A bent strut

35. Camber is found to be out of adjustment on the left front of a vehicle. There is not enough adjustment remaining to correct the alignment. Any of these could be the cause EXCEPT:

    A. A bent control arm.
    B. Incorrect ride height.
    C. Incorrect thrust angle.
    D. A bent spindle.

36. Excessive rear positive toe is found on both sides of a front-wheel drive vehicle with a solid rear axle. Which of the following could be the cause?

    A. A bent front axle
    B. A bent rear axle
    C. A bent lower control arm on the rear
    D. A bent upper control arm on the rear

37. During an alignment, the technician finds caster to be below specification. Technician A says this could be caused by an incorrect rear ride height setting. Technician B says this could be caused by a malfunctioning air suspension system. Who is correct?

    A. A only
    B. B only
    C. Both A and B
    D. Neither A nor B

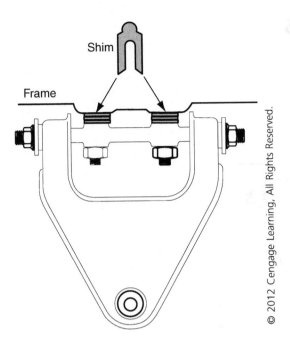

38. Camber needs to be moved toward the positive on the suspension system shown in the figure above. Caster does not need to be changed. Which of the following would be the correct method?

    A. Remove an equal amount of shims from both sides.
    B. Add an equal amount of shims to both sides.
    C. Add shims to the front and remove an equal amount from the back.
    D. Remove shims from the front and add an equal amount to the back.

39. Which of the following best describes positive caster?

    A. The forward tilt of the steering axis
    B. The tilt of the tire inward
    C. The rearward tilt of the steering axis
    D. The tilt of the tire outward

40. The tires on the front of a vehicle are both worn on the inside edges. Technician A says the vehicle may have too much positive toe. Technician B says the vehicle may have too much positive camber. Who is correct?

    A. A only
    B. B only
    C. Both A and B
    D. Neither A nor B

PREPARATION EXAM 5

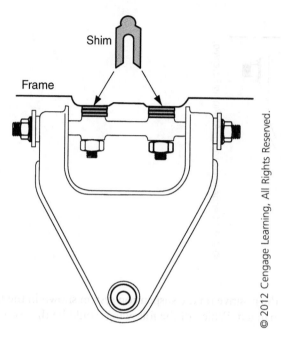

Shim

Frame

1. Camber needs to be moved toward the negative on the suspension system shown in the figure above. Caster does not need to be changed. Which of the following would be the correct method?

   A. Remove an equal amount of shims from both sides.
   B. Add an equal amount of shims to both sides.
   C. Add shims to the front and remove an equal amount from the back.
   D. Remove shims from the front and add an equal amount to the back.

2. Which of the following best describes positive camber?

   A. The forward tilt of the steering axis
   B. The tilt of the tire inward
   C. The rearward tilt of the steering axis
   D. The tilt of the tire outward

3. Which of the following best describes positive toe?

   A. The front of the tires are closer together than the back.
   B. The rear of the tires are closer together than the front.
   C. The left side of the vehicle has a shorter wheelbase than the right.
   D. The right side of the vehicle has a longer wheelbase than the left.

4. Which of the following best describes positive setback?

   A. The front of the tires are closer together than the back.
   B. The rear of the tires are closer together than the front.
   C. The left side of the vehicle has a shorter wheelbase than the right.
   D. The right side of the vehicle has a shorter wheelbase than the left.

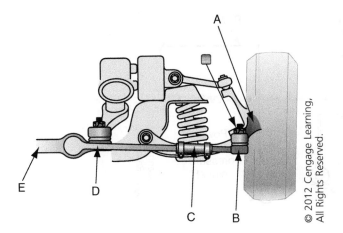

5. Steering axis inclination (SAI) is out of specification on the vehicle shown in the figure above. Which of the following could be the cause?

   A. Item A is bent.
   B. Item B is bent.
   C. Item C is bent.
   D. Item D is bent.

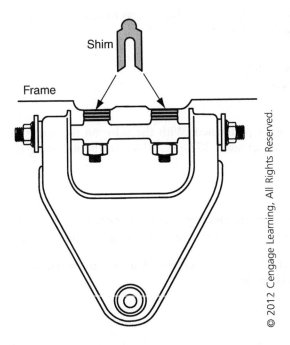

6. Caster needs to be moved toward the negative on the suspension system shown in the figure above. Camber does not need to be changed. Which of the following would be the correct method?

   A. Remove an equal amount of shims from both sides.
   B. Add an equal amount of shims to both sides.
   C. Add shims to the front and remove an equal amount from the back.
   D. Remove shims from the front and add an equal amount to the back.

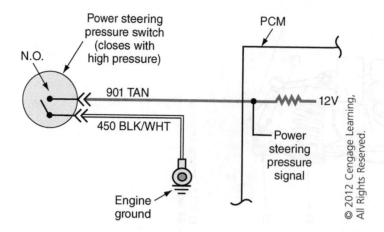

7. A vehicle equipped with the idle compensation system shown in the figure above idles too high at all times. Which of the following could be the cause?

    A. Stuck open pressure switch
    B. Circuit 450 grounded
    C. An open at engine ground
    D. Circuit 901 grounded

8. A customer has an occasional hard steering concern which seems to be more common when it is raining. Which of the following is the most likely cause?

    A. Low fluid level
    B. Faulty power steering pressure switch
    C. Loose power steering pump belt
    D. Low tire pressure

9. A power steering pressure test is being performed. With the valve on the tester closed, the maximum pressure shown on the gauge is 875 psi. Which of the following is indicated?

    A. The pressure is too high.
    B. The pressure is too low.
    C. The pressure is acceptable, however, the steering gear has a restriction.
    D. The pressure is acceptable, however, the steering gear has a leaking internal seal.

10. When a power steering pump pressure test is being performed, what is the maximum amount of time the valve on the tester should be closed?

    A. 5 seconds
    B. 15 seconds
    C. 25 seconds
    D. 35 seconds

11. Consider the following front alignment readings:

	Spec	Actual	
		Left	Right
Caster	0.0 degrees	0.0 degrees	0.0 degrees
Camber	1.5 degrees	1.5 degrees	0.5 degrees
Toe	0.5 degrees	0.2 degrees	0.2 degrees
SAI	2.3 degrees	2.2 degrees	2.4 degrees

Which of the following is a true statement?

A. The technician should adjust the camber, then the toe.
B. The technician should check for a bent steering knuckle.
C. The technician should adjust the toe, then camber.
D. The technician should adjust the caster, then camber.

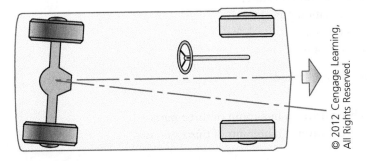

12. The condition shown in the figure above is:

A. Positive thrust.
B. Negative thrust.
C. Positive camber.
D. Negative camber.

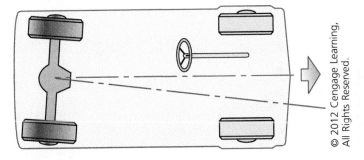

13. The condition shown in the figure above is corrected by:

A. Adjusting rear toe.
B. Adjusting front toe.
C. Adjusting rear camber.
D. Adjusting front camber.

14. Which of the following is the normal tire rotational pattern for all-wheel drive vehicles?

A. Full "X"
B. Modified "X"
C. Front to rear
D. Cross the front, bring to the rear, and send the rears straight forward

15. Which of the following is the normal tire rotational pattern for a front-wheel drive vehicle?
    A. Full "X"
    B. Modified "X"
    C. Front to rear
    D. Cross the front, bring to the rear, and send the rears straight forward

16. When checking the tire pressure on a vehicle, where would a technician find the proper air pressure specification?
    A. In the owner's manual
    B. On an under-hood sticker
    C. On an in-trunk sticker
    D. On the driver's door jamb

17. The tire pressure specification printed on the sidewall of the tire is the:
    A. Recommended cold inflation pressure.
    B. Recommended hot inflation pressure.
    C. Maximum inflation pressure.
    D. Minimum inflation pressure.

18. When should tire pressure be checked?
    A. On a hot tire
    B. After the tire has been driven a minimum of three miles
    C. After the tire has been driven a minimum of three minutes
    D. On a cold tire

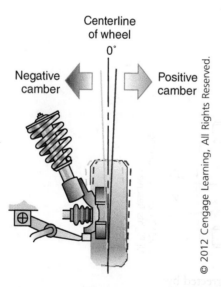

19. Camber needs to be adjusted on the suspension shown in the figure above. Technician A says this can be adjusted by installing new sway bar bushings. Technician B says this can be adjusted by installing new coil springs. Who is correct?
    A. A only
    B. B only
    C. Both A and B
    D. Neither A nor B

20. A vehicle is hard to steer. Any of these could be the cause EXCEPT:

  A. Seized king pins.
  B. Seized ball joints.
  C. Loose wheel bearings.
  D. Excessive positive caster.

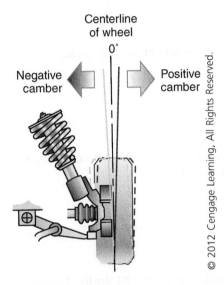

Centerline of wheel

0°

Negative camber   Positive camber

© 2012 Cengage Learning, All Rights Reserved.

21. The lower ball joint is being replaced on the suspension system shown in the figure above. Which of the following is correct?

  A. The jack stand must be placed under the lower control arm.
  B. The drive axle must be removed.
  C. The strut must be removed from the vehicle.
  D. The upper strut mount must be removed.

22. A vehicle pulls to the right. Technician A says the power steering pump could be the cause. Technician B says the power steering gear could be the cause. Who is correct?

  A. A only
  B. B only
  C. Both A and B
  D. Neither A nor B

23. The EPS light is continuously illuminated on the dash. What should the technician do?

  A. Connect a scan tool and access the trouble code.
  B. Align the front axle.
  C. Check the power steering fluid level.
  D. Center the steering wheel.

24. An icon of a steering wheel is illuminated on the dash. Which of the following is indicated?

  A. The power steering fluid is low.
  B. The power steering pressure is low.
  C. There is a problem with the tire pressure monitoring system.
  D. There is a problem in the EPS system.

25. Technician A says that when reinstalling the clock spring, the steering needs to be in a straight ahead position. Technician B says that when installing a clock spring, the clock spring needs to be centered. Who is correct?

   A. A only
   B. B only
   C. Both A and B
   D. Neither A nor B

26. The center link on a steering system is unlevel. This would most likely cause:

   A. A pull to the left.
   B. A pull to the right.
   C. Bump steer.
   D. Incorrect thrust angle.

27. Technician A says a pitman arm should be removed using heat. Technician B says a pitman arm should be removed using a puller. Who is correct?

   A. A only
   B. B only
   C. Both A and B
   D. Neither A nor B

28. A vehicle steering wheel is crooked with the vehicle sitting still and the front wheels straight ahead. Which of the following is the most likely cause?

   A. A bent pitman arm
   B. Incorrect thrust line setting
   C. Incorrect thrust angle setting
   D. A bent rear axle

29. Which of the following could be changed if a vehicle had a bent rear axle?

   A. Front camber
   B. Front toe
   C. Front camber
   D. Thrust angle

30. A vehicle with four-wheel steering is being aligned. Which of the following is true?

   A. The steering wheel must be locked before the rear toe can be adjusted.
   B. The rear steering must be locked before the rear toe is adjusted.
   C. The front camber must be set before the rear toe.
   D. The front toe must be set before rear toe.

31. During a pre-alignment inspection, the technician finds rust streaks on one rear leaf spring. Which of the following should be performed?

   A. The spring should be primed, then painted.
   B. The spring should be replaced.
   C. Both rear springs should be replaced.
   D. The spring should be inspected for cracks.

32. There are rust streaks on the rear leaf spring where it contacts the rear axle. Which of the following is true?

    A. The spring must be replaced.
    B. The spring center bolt may be sheared.
    C. The rear camber needs to be adjusted.
    D. The rear axle must be replaced.

33. An upper strut mount needs to be replaced on a double wishbone front suspension. Technician A says the strut must also be replaced. Technician B says the coil spring must also be replaced. Who is correct?

    A. A only
    B. B only
    C. Both A and B
    D. Neither A nor B

34. A vehicle with torsion bar front suspension is low on the left front corner. Which of the following is the correct repair procedure?

    A. Attempt to adjust the right front torsion bar.
    B. Attempt to adjust the left front torsion bar.
    C. Replace the left front torsion bar.
    D. Replace both torsion bars.

35. The torsion bars on a vehicle have been removed and are being reinstalled. Which of the following is correct concerning reinstallation?

    A. The torsions bars must be replaced.
    B. The torsion bars should be reinstalled on the opposite side from removal to equalize wear.
    C. The torsion bars should be turned end-for-end to equalize wear.
    D. The final adjustment should be in the upward direction.

36. A right rear lower ball joint is being replaced on an independent rear suspension. Where should the jack stand be placed during this procedure?

    A. Under the frame on the right side
    B. Under the right rear lower control arm
    C. Under the frame on the left side
    D. Under the left rear control arm

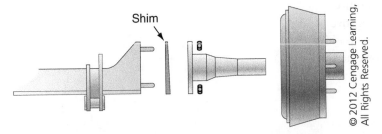

37. A vehicle equipped with the suspension system shown in the figure above is having the spindle replaced. The shim being installed as shown above will:

    A. Correct axle length.
    B. Correct spindle length.
    C. Correct toe.
    D. Correct camber.

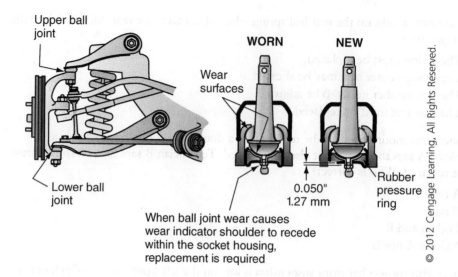

Upper ball joint

Lower ball joint

WORN   NEW

Wear surfaces

0.050"
1.27 mm

Rubber pressure ring

When ball joint wear causes wear indicator shoulder to recede within the socket housing, replacement is required

38. To inspect the lower ball joint on the suspension system shown in the figure above, the technician should:

 A. Place a jack stand under the frame.
 B. Place a jack stand under the lower control arm.
 C. Support the vehicle on a frame hoist.
 D. Inspect the joints with the vehicle setting on the tires.

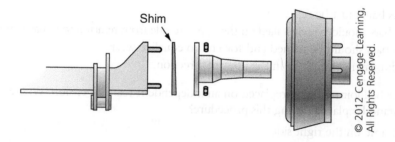

Shim

39. The shim is being installed as shown in the figure above. This will cause which of the following?

 A. Toe to move positive
 B. Toe to move negative
 C. Camber to move positive
 D. Camber to move negative

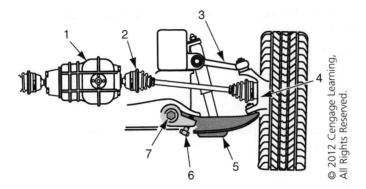

40. Item #6 in the figure above is used to:

    A.   Adjust camber.
    B.   Adjust caster.
    C.   Adjust toe.
    D.   Adjust ride height.

PREPARATION EXAM 6

1. When should the power steering fluid be checked?

    A.   Only when the fluid is hot
    B.   Only when the fluid is cold
    C.   When the fluid is hot or cold
    D.   When the steering is straight ahead

2. A power steering pump makes noise all the time. Which of the following is the most likely cause?

    A.   Aerated fluid
    B.   Stuck open pressure relief valve
    C.   Stuck closed pressure relief valve
    D.   Slightly loose belt

3. Which of the following is the most likely cause of a tire which is worn on both outside shoulders?

    A.   Underinflation
    B.   Overinflation
    C.   Positive caster
    D.   Positive camber

4. A tire has a leak in the sidewall. Which of the following is the correct repair procedure?

    A.   Locate the source of the leak and plug it from the outside.
    B.   Locate the source of the leak and plug it from the inside.
    C.   Locate the source of the leak, plug it from the outside, and patch it from the inside.
    D.   Replace the tire.

5. A customer is concerned that it takes increasing effort to steer the car. Which of the following could be the cause?
   A. Worn steering column bushings
   B. Seized steering column U-joint
   C. Worn tie rod end
   D. Loose idler arm

6. A car equipped with power steering pulls (leads) to the left after a proper alignment. Which of the following is the most likely cause?
   A. A power steering gear
   B. Power steering pump
   C. Pitman arm
   D. All four tires overinflated

7. During a dry park inspection, the technician finds power steering fluid leaking from the rack and pinion bellows. Which of the following is the most likely cause?
   A. An incorrectly installed bellows
   B. A missing bellows clamp
   C. Excessive power steering pump pressure
   D. Worn rack seals

8. Which of the following is the correct inspection procedure for an idler arm?
   A. With the wheels on the ground, grasp the idler arm and attempt to move it with hand pressure only.
   B. With the wheels on the ground, use a large pair of pliers to attempt to move the idler arm.
   C. Lift the vehicle on a frame hoist, grasp the idler arm, and attempt to move it with hand pressure only.
   D. Lift the vehicle on a frame hoist and turn the steering from side to side looking for play.

9. A vehicle has steering wheel shimmy after crossing a railroad track. Which of the following is the most likely cause?
   A. Worn tie rod end
   B. Worn ball joint
   C. Weak front springs
   D. Worn steering damper

10. A vehicle has worn front and rear jounce bumpers. All of these could be the cause EXCEPT:
    A. Worn shocks.
    B. Weak springs.
    C. Worn wheel bearings.
    D. Incorrect ride height.

11. Both front springs have been replaced on the vehicle. Which of the following must also be performed?
    A. Align the vehicle
    B. Replace the rear springs
    C. Replace the shocks
    D. Perform a power steering pressure test

12. A strut bearing is noisy and must be replaced. Technician A says to replace the strut cartridge while replacing the strut bearing. Technician B says to replace the front coil spring while replacing the strut bearing. Who is correct?

    A. A only

    B. B only

    C. Both A and B

    D. Neither A nor B

13. A vehicle sits lower on the left rear than on the right rear. Which of the following is the most likely cause?

    A. The left rear shock is weak.

    B. The right shock was replaced and the left shock was not.

    C. The lateral link bushings are worn.

    D. The left rear spring is weak.

14. A vehicle has worn sway bar end link bushings. Which of the following would be the most likely customer complaint?

    A. Noise while driving in a straight ahead position on a smooth road

    B. Noise while driving on an uneven road

    C. Lower than normal ride height

    D. Tire wear on the outside of the tread

15. A composite leaf spring is being replaced on a vehicle. Which of the following is a true statement about this procedure?

    A. The spring should be lubricated prior to installation.

    B. The spring must first be compressed in a shop press.

    C. Care should be taken not to scratch the spring.

    D. The spring can be installed with either side up.

16. A vehicle equipped with air ride rear suspension will not lower when a load is removed from the trunk. Which of the following could be the cause?

    A. A plugged air supply solenoid

    B. A worn air compressor

    C. A faulty air compressor relay

    D. A plugged air vent solenoid

17. Vehicle ride height is correct on the front, but lower than specification on the rear. Which of the following front alignment angles would be most affected?

    A. Toe

    B. Caster

    C. Camber

    D. SAI

18. Which alignment angle is adjusted last during a four-wheel alignment?

    A. Front toe

    B. Rear toe

    C. Front camber

    D. Rear camber

19. Toe-out-on-turns is not within specification. Which of the following is the most likely cause?

    A. Bent spindle

    B. Bent steering arm

    C. Bent strut

    D. Bent shock

20. Technician A says rear toe on some vehicles can be adjusted with shims. Technician B says rear toe on some vehicles can be adjusted with cams. Who is correct?

    A. A only
    B. B only
    C. Both A and B
    D. Neither A nor B

21. A four-wheel alignment is being performed on a vehicle. The thrust angle needs to be corrected. Which adjustment is made to adjust the thrust angle?

    A. Rear camber
    B. Rear toe
    C. Front camber
    D. Front toe

22. A tire has the following marked on the sidewall: P235/75R16. Which of the following is true?

    A. The tire has a section width of 75 mm.
    B. The tire has an aspect ratio of 235.
    C. The tire has been re-grooved.
    D. The tire fits a 16 inch diameter rim.

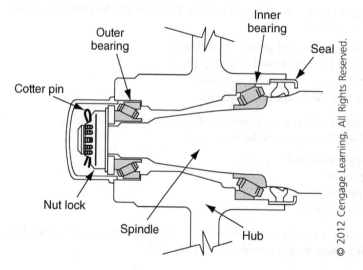

23. The wheel bearings shown in the figure above are being serviced. Which of the following is true concerning the correct tightening procedure?

    A. Tighten the nut to 100 ft lbs.
    B. Tighten the nut to specification and back off two turns.
    C. The final check should be made with a dial indicator.
    D. The wheels/tires must be on the ground.

24. A power steering pump makes noise while turning the steering wheel. Which of the following is the most likely cause?

    A. Aerated fluid
    B. Stuck open pressure relief valve
    C. Stuck closed pressure relief valve
    D. Slightly loose belt

25. Which of the following is the most likely cause of tires which are worn in the center?

    A. Underinflation
    B. Overinflation
    C. Positive caster
    D. Positive camber

26. After setting toe to manufacture specification, the technician checks toe-out-on-turns. When the front left wheel is turned out 20 degrees, the right front wheel turns in 24 degrees. What does this indicate?

    A. A normal operating steering system
    B. A bent strut
    C. A bent steering arm
    D. A worn ball joint

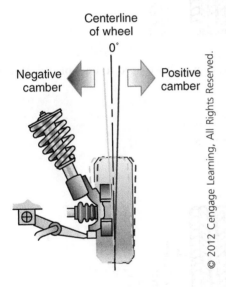

27. The suspension system shown in the figure above is being inspected. Technician A says the load carrying joint is the lower ball joint. Technician B says the load carry joint is the upper strut mount. Who is correct?

    A. A only
    B. B only
    C. Both A and B
    D. Neither A nor B

28. A vehicle with conventional (recirculating ball) steering has a diagnostic code for a faulty steering wheel position sensor. Where is the most likely location for the sensor?
    A. At the base of the steering column
    B. Behind the steering wheel
    C. On the steering gear box
    D. Inside the steering gear box

29. When reinstalling the steering wheel position sensor:
    A. The vehicle alignment must be adjusted.
    B. The sensor may need to be centered.
    C. The power steering system will need to be bled.
    D. The power steering pressure will need to be checked.

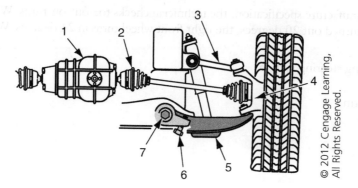

30. The pivot bar bushings in item #3 shown in the figure above need to be replaced. Which of the following is correct?
    A. Item #2 must be removed.
    B. Item #5 must be removed.
    C. The jack stand should be placed under item #5.
    D. Item #6 should be removed.

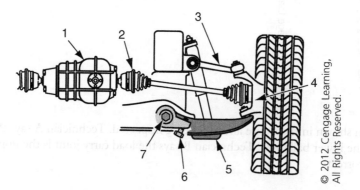

31. Camber needs to be adjusted on the suspension system shown in the figure above. This would most likely occur at:
    A. #3.
    B. #4.
    C. #5.
    D. #6.

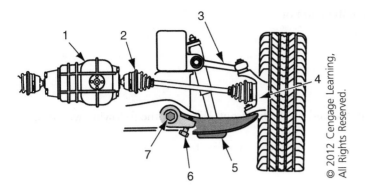

32. Technician A says the suspension system shown in the figure above is a McPherson strut. Technician B says the suspension system shown in the figure above is adjustable. Who is correct?

   A. A only
   B. B only
   C. Both A and B
   D. Neither A nor B

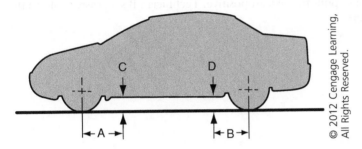

33. In the figure shown above, dimension D is less than specification. Dimension C is within specification. Which of the following is the correct repair procedure?

   A. Replace all 4 springs.
   B. Replace the front two springs.
   C. Replace the rear two springs.
   D. Replace the springs on the left side of the vehicle.

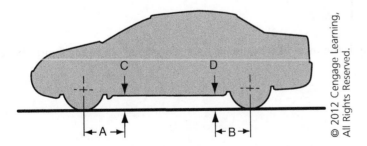

34. In the figure shown above, dimensions C and D are greater than specification. Technician A says all four springs must be replaced. Technician B says this could be caused by using aftermarket tires and wheels. Who is correct?

   A. A only
   B. B only
   C. Both A and B
   D. Neither A nor B

35.   Included angle is the combination of:

A.   Camber and caster.
B.   Front and rear toe.
C.   SAI and camber.
D.   SAI and caster.

36.   A vehicle has a worn tie rod end on the left front. Which of the following would be affected?

A.   Rear toe
B.   Front camber
C.   Front toe
D.   Front caster

37.   A vehicle has a worn tie rod end on the left rear. Which of the following would be affected?

A.   Rear toe
B.   Rear camber
C.   Front toe
D.   Front caster

38.   Technician A says camber pulls to the least positive. Technician B says caster pulls to the most positive. Who is correct?

A.   A only
B.   B only
C.   Both A and B
D.   Neither A nor B

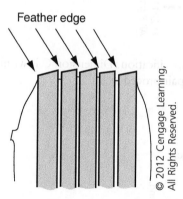

Feather edge

© 2012 Cengage Learning, All Rights Reserved.

39.   Which of the following is the most likely cause of the tire wear shown in the figure above?

A.   Toe
B.   Camber
C.   Caster
D.   SAI

40.   The center link needs to be replaced on a vehicle. Technician A says the outer tie rods will also need to be replaced. Technician B says the toe will need to be adjusted. Who is correct?

A.   A only
B.   B only
C.   Both A and B
D.   Neither A nor B

Answer Keys and Explanations

INTRODUCTION

Included in this section are the answer keys for each preparation exam, followed by individual, detailed answer explanations and a reference identifying the designated task area being assessed by each specific question. This additional reference information may prove useful if you need to refer back to the Task List located in section 4 of this book for additional support.

PREPARATION EXAM 1—ANSWER KEY

1.	C	21.	B
2.	C	22.	C
3.	D	23.	B
4.	A	24.	B
5.	C	25.	B
6.	B	26.	B
7.	B	27.	B
8.	D	28.	A
9.	A	29.	C
10.	A	30.	A
11.	C	31.	C
12.	C	32.	A
13.	C	33.	B
14.	D	34.	B
15.	A	35.	C
16.	D	36.	C
17.	A	37.	D
18.	D	38.	C
19.	A	39.	B
20.	B	40.	D

PREPARATION EXAM 1—EXPLANATIONS

TASK A.2.13

1. Which of the following is the proper way to adjust the rack yoke bearing?

   A. Tighten to 75 ft lbs.

   B. Tighten to 5 inch lbs.

   C. Tighten and then back off the original equipment manufacturer's (OEM) specifications.

   D. Tighten to 0.002" end-play.

   Answer A is incorrect. 75 ft lbs would be too tight.

   Answer B is incorrect. 5 inch lbs would be too loose.

   Answer C is correct. Many OEMs will recommend that the technician tighten to a specification then back off a certain number of degrees.

   Answer D is incorrect. The rack yoke is not adjusted to a clearance specification. There is no way to effectively measure the clearance on this type of adjustment, and the yoke needs to be preloaded with no clearance.

TASK A.3.2

2. A customer says that the steering wheel turns more turns to the left than the right. Which of the following is the LEAST LIKELY cause?

   A. Incorrectly timed steering gear

   B. Bent pitman arm

   C. Faulty power steering pump

   D. Bent tie rod

   Answer A is incorrect. An incorrectly timed steering gear box would cause the box to be off center in the straight ahead position, which would cause this condition.

   Answer B is incorrect. A bent pitman arm would cause the steering box to be off center in the straight ahead position which would cause this condition.

   Answer C is correct. A faulty power steering pump would not change the relationship between the steering wheel and the steering stops.

   Answer D is incorrect. A bent tie rod end would cause the steering to be off center, thus turning further in one direction than the other.

TASK A.1.2

3. A vehicle is hard to steer in one direction only. Which of the following could be the cause?

   A. Tight joint in the steering shaft

   B. Loose joint in the steering shaft

   C. Faulty power steering pump

   D. Faulty power steering gear

   Answer A is incorrect. A tight joint in the steering shaft would cause tight steering either left or right.

   Answer B is incorrect. A loose joint in the steering shaft would cause loose steering in both directions.

   Answer C is incorrect. A faulty power steering pump would affect steering in both directions.

   Answer D is correct. A leaking seal in a power steering gear could affect steering assist in only one direction.

4. The rear tires have a diagonal wear pattern. Which of the following is the most likely cause?

   A. Rear toe
   B. Rear camber
   C. Front camber
   D. Front caster

TASK D.1

Answer A is correct. Incorrect rear toe causes a diagonal swipe wear pattern on the rear tires.

Answer B is incorrect. Rear camber causes shoulder wear on the rear tires.

Answer C is incorrect. Front camber causes shoulder wear on the front tires.

Answer D is incorrect. Front caster is not a tire wear angle.

5. A customer says their car makes more noise on asphalt than on concrete road surfaces. Which of the following could be the cause?

   A. Front wheel bearing wear
   B. Rear wheel bearing wear
   C. Aggressive tire tread patterns
   D. Incorrect alignment angles

TASK D.2

Answer A is incorrect. Wheel bearing noises will change with load, not road surface finish.

Answer B is incorrect. Wheel bearing noises will change with load, not road surface finish.

Answer C is correct. Tires with very aggressive tread patterns are noisy. This noise will vary with different types of finish on the road surface.

Answer D is incorrect. Alignment angles will not cause tire noise to change while driving.

6. There is a loud roaring noise while driving. The noise gets louder when vehicle speed increases. Which of the following could be the cause?

   A. Worn power steering pump
   B. Worn wheel bearing
   C. Worn ball joint
   D. Worn power steering gear

TASK D.2

Answer A is incorrect. Power steering pump noise will vary with engine speed and power steering load, but is not typically vehicle speed sensitive.

Answer B is correct. This is a typical wheel bearing diagnosis.

Answer C is incorrect. A worn ball joint can cause a knocking noise in certain suspension travel positions. However, it would not cause a roaring noise.

Answer D is incorrect. A worn power steering gear does not normally create noise. However, if it did, it would not be vehicle speed dependent.

TASK C.14

7. A setback condition is found during an alignment. Which of the following could be the cause?

  A. Worn spring shackle

  B. Worn spring eye bushing

  C. Worn tie rod end

  D. Worn pitman arm

Answer A is incorrect. The spring shackle is used on the rear of the leaf spring. This would not cause setback.

Answer B is correct. A worn spring eye bushing could allow the rear axle to shift and setback to occur.

Answer C is incorrect. A worn tie rod will affect the steering, not setback.

Answer D is incorrect. A worn pitman arm will affect the steering, not setback.

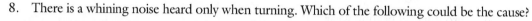

TASK A.2.5

8. There is a whining noise heard only when turning. Which of the following could be the cause?

  A. Leaking internal seals in the steering gear

  B. Worn ball joint

  C. Worn power steering gear

  D. Worn power steering pump

Answer A is incorrect. Leaking internal seals in the steering gear will cause low steering assist.

Answer B is incorrect. A worn ball joint will not cause a whining noise.

Answer C is incorrect. A worn power steering gear will cause low power steering assist.

Answer D is correct. A worn power steering pump can cause noise while turning.

TASK B.1.9

9. The front leaf springs have been replaced on a vehicle. Which of the following operations must also be performed?

  A. Vehicle alignment

  B. Tire rotation

  C. Tire balance

  D. Rear spring replacement

Answer A is correct. After the springs are replaced, the vehicle alignment will be changed. Therefore, the vehicle must be aligned.

Answer B is incorrect. The tires do not have to be rotated after spring replacement.

Answer C is incorrect. The tires do not have to be balanced after spring replacement.

Answer D is incorrect. It is necessary to replace the springs across an axle in pairs to prevent leaning. It is not necessary to replace the springs on both ends of the vehicle at the same time.

10. A vehicle has worn jounce bumpers. Which of the following could be the cause?

    A. Leaking shocks
    B. Leaking power steering pump
    C. Leaking power steering gear
    D. Worn strut bearings

TASK B.2.5

    Answer A is correct. Leaking shocks can allow the vehicle to bottom out on rough road surfaces.

    Answer B is incorrect. A leaking power steering pump would not change suspension travel.

    Answer C is incorrect. A leaking power steering gear would not change suspension travel.

    Answer D is incorrect. Worn strut bearings would not change suspension travel.

11. A worn idler arm would affect which alignment angle?

    A. Caster
    B. Camber
    C. Toe
    D. SAI

TASK A.3.4

    Answer A is incorrect. Caster is not affected by the idler arm.

    Answer B is incorrect. Camber is not affected by the idler arm.

    Answer C is correct. Toe is directly affected by the idler arm. A worn idler arm would allow toe to change while traveling down the road.

    Answer D is incorrect. SAI would not change due to a worn idler arm.

12. A shifted front cradle could affect any of these EXCEPT:

    A. Front camber.
    B. Front caster.
    C. Thrust line.
    D. Thrust angle.

TASK C.15

    Answer A is incorrect. Front camber would be affected.

    Answer B is incorrect. Front caster could be affected.

    Answer C is correct. The thrust line is the direction the rear wheels are pointing; this would not be affected by a shifted front cradle.

    Answer D is incorrect. The thrust angle would be affected because the angle is formed by the thrust line and the geometric center line. The geometric center line is a line drawn through the center points of the front and rear axle. The geometric center line can be changed when the front cradle is shifted.

TASK B.1.1

13. There is a pull on acceleration and deceleration. Technician A says spring eye bushings could be the cause. Technician B says worn strut rod bushings could be the cause. Who is right?

    A. A only
    B. B only
    C. Both A and B
    D. Neither A nor B

    Answer A is incorrect. Technician B is also correct.

    Answer B is incorrect. Technician A is also correct.

    Answer C is correct. Both Technicians are correct. Spring eye bushings and worn strut rod bushings can both allow the control arm to move when the vehicle is driven.

    Answer D is incorrect. Both Technicians are correct.

TASK B.2.4

14. A composite leaf spring is being installed in a vehicle. Which of the following is correct?

    A. The spring can be installed using a pry bar.
    B. The spring can be compressed with vise grips.
    C. The spring should be painted after installation.
    D. The spring should not be scratched during installation.

    Answer A is incorrect. The spring should not be pried on with a pry bar.

    Answer B is incorrect. The spring should not have vise grips applied.

    Answer C is incorrect. The spring should not be painted.

    Answer D is correct. The spring needs to be handled with care. If the spring is dented or scratched, it will be weakened and can break.

TASK A.3.6

15. Which of the following alignment specifications would most likely need a steering damper?

    A. 6 degrees caster
    B. 2 degrees camber
    C. 2 degrees caster
    D. 6 degrees camber

    Answer A is correct. Vehicles with high caster settings are most likely to use a steering damper to help control front-wheel shimmy.

    Answer B is incorrect. Vehicles with high caster settings are most likely to use a steering damper to help control front-wheel shimmy.

    Answer C is incorrect. Vehicles with high caster settings are most likely to use a steering damper to help control front-wheel shimmy.

    Answer D is incorrect. Vehicles with high caster settings are most likely to use a steering damper to help control front-wheel shimmy.

16. A customer complains of a popping noise while turning the steering wheel. Which of the following is the most likely cause?

    A.   Rear shock mounts
    B.   Rear toe adjusters
    C.   Front shock mounts
    D.   Front tie rods

TASK A.2.1

    Answer A is incorrect. Worn rear shock mounts can cause noise while traveling over bumps.

    Answer B is incorrect. Worn rear toe adjusters do not normally make noise.

    Answer C is incorrect. Worn front shock mounts can make noise when traveling over bumps.

    Answer D is correct. Worn front tie rods can make a popping noise when the steering wheel is rocked back and forth.

17. A power steering pump hose is leaking on a vehicle. Which of the following is the most likely cause?

    A.   Separated motor mounts
    B.   Loose drive belt
    C.   Air in the fluid
    D.   Broken pump mounts

TASK A.2.5

    Answer A is correct. It is possible that separated motor mounts could cause the hoses to be stretched and leak.

    Answer B is incorrect. A loose drive belt will not cause the hoses to leak.

    Answer C is incorrect. Air in the fluid will cause noise but will not cause the hose to leak.

    Answer D is incorrect. Broken pump mounts would not allow the rack to move enough to cause the hose to leak.

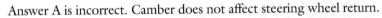

18. Which angle has the greatest affect on steering wheel return?

    A.   Camber
    B.   SAI
    C.   Toe
    D.   Caster

TASK B.1.1

    Answer A is incorrect. Camber does not affect steering wheel return.

    Answer B is incorrect. SAI does not affect steering wheel return.

    Answer C is incorrect. Toe does not affect steering wheel return.

    Answer D is correct. Caster greatly affects steering wheel return.

TASK D.4

19. There is a vibration which can be felt in the steering wheel at speeds above 30 mph. Which of the following could be the cause?

   A. Front wheels out of balance
   B. Rear wheels out of balance
   C. Caster
   D. Camber

   Answer A is correct. A vibration felt in the steering wheel is most likely caused by the front wheels being out of balance.

   Answer B is incorrect. Rear wheels out of balance would cause vibrations normally felt in the seat.

   Answer C is incorrect. Caster is not associated with vibrations.

   Answer D is incorrect. Camber is not associated with vibrations.

TASK B.1.7

20. The hole in the steering knuckle into which the ball joint stud goes is out of round. Which of the following is the best method of repair?

   A. Measure the wear and check against manufacturer's specifications.
   B. Replace the steering knuckle.
   C. Ream the hole.
   D. Shim the hole.

   Answer A is incorrect. The only acceptable repair method is to replace the steering knuckle.

   Answer B is correct. The steering knuckle must be replaced.

   Answer C is incorrect. Reaming the hole is not an acceptable repair.

   Answer D is incorrect. The hole cannot be shimmed.

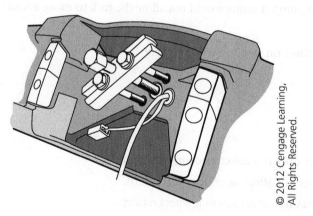

© 2012 Cengage Learning, All Rights Reserved.

TASK A.1.2

21. The setup shown in the photo above is used to:

   A. Remove the horn.
   B. Remove the steering wheel.
   C. Install the air bag.
   D. Install the steering wheel.

   Answer A is incorrect. The tool shown is setup to remove the steering wheel.

   Answer B is correct. The tool shown is setup to remove the steering wheel.

   Answer C is incorrect. The tool shown is a puller, not used for an installation.

   Answer D is incorrect. The tool shown is a puller, not used for an installation.

22. The left front tire on a vehicle shows outside tread wear. Which of the following is the most likely cause?

    A. Worn shock absorbers
    B. Tires overinflated
    C. Camber adjustment incorrect
    D. Caster adjustment incorrect

    TASK D.1

    Answer A is incorrect. Worn shock absorbers will cause cupping across the tire tread.

    Answer B is incorrect. Overinflated tires will cause tread wear in the center of the tread, not outside tread wear.

    Answer C is correct. Excessive positive camber can cause outside tread wear.

    Answer D is incorrect. Caster is not a tire wear angle.

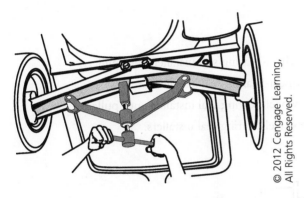

© 2012 Cengage Learning, All Rights Reserved.

23. What is being performed in the photo above?

    A. Installing a longitudinally mounted leaf spring
    B. Removing a transversely mounted leaf spring
    C. Installing a coil spring
    D. Removing a coil spring

    TASK B.2.4

    Answer A is incorrect. This is a transversely mounted leaf spring, not longitudinally.

    Answer B is correct. The technician is using a special tool to compress and remove a transversely mounted leaf spring.

    Answer C is incorrect. This is not a coil spring.

    Answer D is incorrect. This is not a coil spring.

TASK B.1.7

24. The tools shown in the photo above are used to:

A. Remove seals.

B. Install seals.

C. Remove bearings.

D. Install bearings.

Answer A is incorrect. These are wheel-bearing seal installers, not pullers.

Answer B is correct. These are wheel-bearing seal installers.

Answer C is incorrect. These are wheel-bearing seal installers, not pullers.

Answer D is incorrect. These are wheel-bearing seal installers.

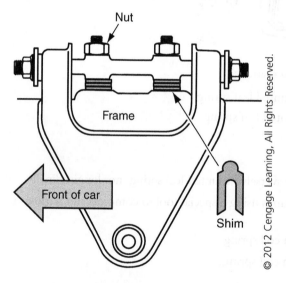

TASK C.6

25. Referring to the figure above, if shims are removed from the rear shim pack and installed in the front spring pack, which of the following will occur?

A. Toe will go positive.

B. Caster will go negative.

C. Caster will go positive.

D. Toe will go negative.

Answer A is incorrect. This is not a toe adjustment.

Answer B is correct. Caster will move toward the front of the car, which is negative.

Answer C is incorrect. Caster will go negative.

Answer D is incorrect. This is not a toe adjustment.

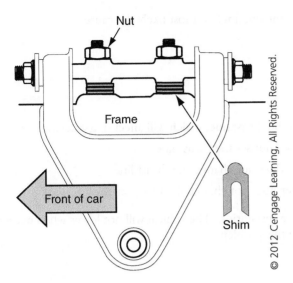

Nut

Frame

Front of car

Shim

26. Referring to the figure above, if shims are added to both shim packs, which of the following will occur?

    A. Toe will go positive.
    B. Camber will go negative.
    C. Camber will go positive.
    D. Toe will go negative.

TASK C.4

Answer A is incorrect. This is not a toe adjustment.

Answer B is correct. Camber will move toward the center of the car, which is negative.

Answer C is incorrect. Camber will go negative.

Answer D is incorrect. This is not a toe adjustment.

27. Which of the following would cause toe-out-on-turns to be out of specification?

    A. Bent tie rod end
    B. Bent steering arm
    C. Bent lower control arm
    D. Bent upper control arm

TASK C.9

Answer A is incorrect. A bent tie rod end would affect toe.

Answer B is correct. A bent steering arm would change toe-out-on-turns.

Answer C is incorrect. A bent lower control arm would change caster and camber.

Answer D is incorrect. A bent upper control arm would change camber and caster.

TASK C.13

28. A vehicle is dog-tracking. Which alignment angle is most likely the cause?

    A. Rear toe
    B. Front toe
    C. Rear camber
    D. Front camber

    Answer A is correct. Rear toe sets the thrust line, which will affect dog-tracking.

    Answer B is incorrect. Front toe does not set the thrust line.

    Answer C is incorrect. Rear camber does not change the thrust line

    Answer D is incorrect. Front camber does not change the thrust line.

TASK C.3

29. A vehicle is equipped with an air ride suspension. The vehicle will not lower when necessary. Which of the following could cause this problem?

    A. Compressor relay
    B. Compressor
    C. Sensor
    D. Fuse

    Answer A is incorrect. The compressor relay is not used to lower the suspension.

    Answer B is incorrect. The compressor is not used to lower the suspension.

    Answer C is correct. The sensor is responsible for sending a signal that the rear is too high.

    Answer D is incorrect. A fuse would most likely affect more than just the lowering of the ride height.

TASK B.2.11

30. Which of the following is the correct way to adjust tapered roller bearings?

    A. Torque and then back off OEM specification.
    B. Torque to OEM specification.
    C. Adjust to a minimum of .150" (3.80 mm) end-play.
    D. Torque to 50 ft lbs.

    Answer A is correct. It is typical to torque the bearing while rotating the wheel and then back off the nut a certain amount.

    Answer B is incorrect. Usually after torquing the nut will be backed off one flat to allow for the proper setting of end-play.

    Answer C is incorrect. This would be excessive end-play.

    Answer D is incorrect. This would load the bearing too much.

31. A vehicle has a crooked steering wheel when driving down the road. Technician A says this could be caused by failure to install the steering wheel correctly. Technician B says this could be caused by rear toe being out of adjustment. Who is correct?

TASK A.1

   A.   A only
   B.   B only
   C.   Both A and B
   D.   Neither A nor B

Answer A is incorrect. Technician B is also correct.

Answer B is incorrect. Technician A is also correct.

Answer C is correct. Both Technicians are correct. Rear toe being out of proper adjustment will cause a thrust angle to exist which will cause a crooked steering wheel. Also, a steering wheel which was installed one spline off could cause a crooked steering wheel.

Answer D is incorrect. Both Technicians are correct.

32. When does a positive camber angle exist?

TASK C.3

   A.   When the top of the tire leans outward
   B.   When the top of the tire leans inward
   C.   When the upper ball joint is rearward of the lower ball joint as viewed from the side
   D.   When the rear axle is crooked

Answer A is correct. The top of the tire leans out with positive camber.

Answer B is incorrect. The top of the tire leans in with negative camber.

Answer C is incorrect. This describes positive caster.

Answer D is incorrect. This describes a thrust line problem.

33. Why is the camber angle important?

TASK C.3

   A.   Because it directly affects ride stiffness
   B.   Because it is a tire wearing angle and it affects steering control
   C.   Because it can cause vibration
   D.   Because it affects the operation of the brakes

Answer A is incorrect. Camber angle does not directly affect ride stiffness.

Answer B is correct. Camber is a tire wear angle and can cause a pull in the steering.

Answer C is incorrect. Camber does not cause a vibration.

Answer D is incorrect. Camber does not affect brake operation.

34. The caster angle on the left front wheel has been set at 1 degree negative. The right front caster angle has been set at 1 degree positive. What would be the most likely result?

TASK C.5

   A.   The car would pull to the right.
   B.   The car would pull to the left.
   C.   The car would travel straight ahead.
   D.   The left tire would show second rib wear.

Answer A is incorrect. A car will usually pull to the most negative caster setting.

Answer B is correct. A car will usually pull to the most negative caster setting.

Answer C is incorrect. The car would most likely pull with a two degree cross-caster setting.

Answer D is incorrect. Caster is not a tire wear angle.

TASK C.10

35. Technician A says that camber and SAI make up the included angle. Technician B says that the purpose of the turning angle is to reduce tire scuffing during cornering. Who is right?

A. A only
B. B only
C. Both A and B
D. Neither A nor B

Answer A is incorrect. Technician B is also correct.

Answer B is incorrect. Technician A is also correct.

Answer C is correct. Both Technicians are correct. Included angle is found by combining camber and SAI. The purpose of the turn angle is to allow the front tires to travel through a turn in different circles. Tire scuffing during cornering is controlled by toe-out on-turns or turning angle.

Answer D is incorrect. Both Technicians are correct.

TASK C.12

36. Technician A says that shims are sometimes used to adjust rear toe and camber. Technician B says that strut rods are sometimes used to adjust rear toe and camber. Who is right?

A. A only
B. B only
C. Both A and B
D. Neither A nor B

Answer A is incorrect. Technician B is also correct.

Answer B is incorrect. Technician A is also correct.

Answer C is correct. Both Technicians are correct. Shims can be installed behind the rear bearing assembly to change the rear toe. Strut rods can also be used to change rear toe on some models.

Answer D is incorrect. Both Technicians are correct.

TASK C.7

37. When the rear of the rear tires are closer together than at the front, the tires have:

A. Negative camber.
B. Positive camber.
C. Toe-in.
D. Toe-out.

Answer A is incorrect. Camber is the tilt of the tire in or out.

Answer B is incorrect. Camber is the tilt of the tire in or out.

Answer C is incorrect. Toe-in has the front of the tires closer together than the rear.

Answer D is correct. Toe-out has the front of the tires farther apart than the rear.

38. The jounce bumpers on the rear of a vehicle are severely worn. Which of the following is the LEAST LIKELY cause?

    A. Worn rear springs
    B. Excessive cargo in the trunk.
    C. Worn front springs
    D. Worn rear shocks

TASK B.2.5

    Answer A is incorrect. Worn rear springs can cause worn rear jounce bumpers.

    Answer B is incorrect. Excessive weight in the rear of a vehicle can cause the suspension to sag and cause the rear jounce bumpers to be worn excessively.

    Answer C is correct. Worn front springs are the least likely cause of worn rear jounce bumpers.

    Answer D is incorrect. Worn rear shocks can allow increased travel of the rear suspension and wear the jounce bumpers excessively.

39. A customer has a noise concern coming from the rear of the vehicle. Any of these could be the cause EXCEPT:

    A. Worn rear wheel bearings.
    B. Rear tires out of balance.
    C. Missing spring separators.
    D. Worn spring eye bushings.

TASK B.2

    Answer A is incorrect. Worn rear wheel bearings can cause noise from the rear.

    Answer B is correct. Rear tires out of balance would cause a vibration but not a noise.

    Answer C is incorrect. Missing spring separators can cause noisy leaf springs.

    Answer D is incorrect. Worn spring eye bushings can cause a noise from the rear of the vehicle.

40. Which of the following is true concerning torsion bar front suspension?

    A. Torsion bars must always be replaced in pairs.
    B. Torsion bars are non-adjustable.
    C. When torsion bars are reinstalled, they can be installed in either side.
    D. Torsion bars can be used on SLA suspensions.

TASK B.1

    Answer A is incorrect. Because they are adjustable, torsion bars can be replaced individually.

    Answer B is incorrect. Torsions bars are adjustable.

    Answer C is incorrect. When reusing torsion bars, they should be reinstalled in the same position as where they were removed.

    Answer D is correct. SLA suspensions often use torsion bars instead of coil springs.

PREPARATION EXAM 2—ANSWER KEY

1.	C	21.	D
2.	B	22.	C
3.	A	23.	B
4.	D	24.	D
5.	D	25.	B
6.	D	26.	C
7.	C	27.	A
8.	A	28.	B
9.	B	29.	D
10.	C	30.	B
11.	C	31.	A
12.	D	32.	C
13.	C	33.	B
14.	B	34.	A
15.	B	35.	B
16.	D	36.	D
17.	C	37.	A
18.	A	38.	B
19.	A	39.	C
20.	B	40.	B

PREPARATION EXAM 2—EXPLANATIONS

TASK A.1.1

1. Which of the following could cause poor returnability?

    A. Thrust angle
    B. Toe-out-on-turns
    C. Tire pressure
    D. Power steering pump pressure

    Answer A is incorrect. Thrust angle will cause a crooked steering wheel, but it is unlikely to be the cause of poor returnability.

    Answer B is incorrect. Improper toe-out-on-turns will cause tire scuffing in corners, not returnability concerns.

    Answer C is correct. Low tire pressure will cause poor returnability and is a common problem.

    Answer D is incorrect. Power steering pump pressure is unlikely to be the cause.

2. Which of the following alignment angles is most likely to be the cause of a steering returnability concern?

    A. Front toe

    B. Caster

    C. Camber

    D. Rear toe

TASK C.1

Answer A is incorrect. Front toe is a tire wear angle but has little effect on returnability.

Answer B is correct. Caster has the greatest affect on returnability.

Answer C is incorrect. Camber has less affect on returnability than caster.

Answer D is incorrect. Rear toe has less affect on returnability than caster.

3. Technician A says the SAI/KPI (steering axis inclination/key performance indicator) angle can be used to diagnose a bent strut. Technician B says the SAI/KPI angle can be used to diagnose a bent steering arm. Who is correct?

    A. A only

    B. B only

    C. Both A and B

    D. Neither A nor B

TASK B.1.12

Answer A is correct. Only Technician A is correct. SAI/KPI is very useful when trying to identify a bent strut.

Answer B is incorrect. Technician A is correct. A bent steering arm is diagnosed by the turning angle.

Answer C is incorrect. Only Technician A is correct.

Answer D is incorrect. Technician A is correct.

4. Any of these would be an indication of a front cradle out of alignment EXCEPT:

    A. SAI/KPI varies 2 degrees side to side

    B. Camber positive on the left side and negative on the right side

    C. Camber negative on the left side and positive on the right side

    D. Thrust angle of 0.1 degrees

TASK C.15

Answer A is incorrect. SAI/KPI variance of 2 degrees is an excellent indication of a front cradle out of alignment.

Answer B is incorrect. When camber is positive on one side and negative on the other, front cradle alignment is suspect.

Answer C is incorrect. When camber is positive on one side and negative on the other, front cradle alignment is suspect.

Answer D is correct. A thrust angle of 0.1 degrees is well within specifications and would not indicate a misaligned cradle.

TASK C.1

5. The owner of a vehicle equipped with manual steering complains of hard steering. The alignment is checked and the readings are below:

	Spec	Actual Left	Actual Right
Caster	0.0 degrees	2.0 degrees	2.1 degrees
Camber	0.5 degrees	0.6 degrees	0.5 degrees
Toe	0.5 degrees	0.2 degrees	0.2 degrees

Which of the following is the most likely cause?

A. Camber is too negative.
B. Caster is too negative.
C. Camber is too positive.
D. Caster is too positive.

Answer A is incorrect. Camber is close enough to specification to be within tolerance.

Answer B is incorrect. Caster is not too negative on this vehicle.

Answer C is incorrect. Camber is close enough to specification to be within tolerance.

Answer D is correct. Caster is set too positive on this vehicle and is most likely the cause of the hard steering complaint.

TASK C.3

6. Refer to the alignment readings below:

	Spec	Actual Left	Actual Right	Tolerance
Caster	0.0 degrees	0.0 degrees	0.1 degrees	+/– 0.2
Camber	0.5 degrees	0.6 degrees	0.5 degrees	+/– 0.2
Toe	0.4 degrees	0.2 degrees	0.2 degrees	+/– 0.1

Which of the following is a true statement?

A. Caster should be adjusted.
B. Camber should be adjusted.
C. Caster and camber are within tolerance and toe should be adjusted.
D. This vehicle is within alignment tolerances.

Answer A is incorrect. Caster is within tolerance.

Answer B is incorrect. Camber is within tolerance.

Answer C is incorrect. Toe is within tolerance. The spec is for total toe. The actual is individual toe which is half total toe.

Answer D is correct. This vehicle is within alignment tolerances.

7. A vehicle with the following alignment readings is wearing the outside edge of the right front tire. Which of the following is most likely the cause?

TASK D.1

	Spec	Actual		Tolerance
		Left	Right	
Caster	0.0 degrees	2.0 degrees	0.1 degrees	+/− 0.2
Camber	0.5 degrees	0.6 degrees	3.5 degrees	+/− 0.2
Toe	0.5 degrees	0.3 degrees	0.2 degrees	+/− 0.1

A. Left toe
B. Left caster
C. Right camber
D. Right toe

Answer A is incorrect. Left toe is within tolerance.

Answer B is incorrect. Left caster is out of tolerance, however, caster is not a tire wear angle.

Answer C is correct. Right camber is too positive and can cause outside shoulder wear.

Answer D is incorrect. Right toe is within tolerance.

8. A four-wheel drive vehicle with oversized tires and rims has a shimmy after hitting a bump. A dry park inspection has been performed and the alignment has been checked. No problems were identified. Which of the following would most likely fix the concern?

TASK A.3.6

A. Installation of a steering damper
B. Adjust the caster 2 degrees positive from the original caster specifications
C. Adjust the caster 2 degrees negative from the original caster specifications
D. Installation of a power steering cooler

Answer A is correct. Installing a steering damper will help correct a front-wheel shimmy due to oversized tires and wheels.

Answer B is incorrect. Adding caster will only add to the problem.

Answer C is incorrect. Subtracting caster will not fix the concern and may only cause the vehicle to have less directional stability.

Answer D is incorrect. A power steering cooler would not help the shimmy concern.

9. A vehicle has a shimmy when the steering wheel returns to center after completing a turn. Which of the following could be the cause?

TASK C.6

A. Excessive positive camber
B. Excessive positive caster
C. Excessive toe-out
D. Excessive toe-in

Answer A is incorrect. Excessive positive camber can cause a pull and tire wear but will not normally cause a shimmy.

Answer B is correct. Excessive positive caster can cause the front wheels to overshoot the straight ahead position when returning to center and result in a shimmy.

Answer C is incorrect. Excessive toe-out usually causes a wander.

Answer D is incorrect. Excessive toe-in usually causes a darting.

TASK C.7

10. Which alignment angle is usually considered the greatest tire wearing angle?

A. Caster

B. Camber

C. Toe

D. SAI/KPI

Answer A is incorrect. Caster is not a tire wearing angle.

Answer B is incorrect. While camber is a tire wearing angle, it is not considered the greatest tire wearing angle.

Answer C is correct. Toe is the greatest tire wearing angle.

Answer D is incorrect. A vehicle with incorrect SAI/KPI, may wear the tires depending on the cause. However, it is not considered the greatest tire wearing angle.

TASK B.2.5

11. Which of the following would be the LEAST LIKELY cause of worn jounce bumpers?

A. Worn shocks

B. Worn springs

C. Worn ball joints

D. A low ride height setting

Answer A is incorrect. Worn shocks can cause the jounce bumpers to be worn.

Answer B is incorrect. Worn springs can cause the jounce bumpers to be worn.

Answer C is correct. Worn ball joints will not cause worn jounce bumpers.

Answer D is incorrect. A low ride height setting can cause the jounce bumpers to wear.

TASK B.2.4

12. A new composite leaf spring is being installed. Which of the following is true?

A. The spring must be coated with a penetrant prior to installation.

B. The spring must be steam cleaned prior to installation.

C. A ball joint separator must be used to slide the spring in to position.

D. A spring compressor must be used to install the spring.

Answer A is incorrect. A new composite spring should be installed dry and clean.

Answer B is incorrect. The spring should not be subjected to the heat and pressure of a steam cleaner.

Answer C is incorrect. A ball joint separator may need to be used to separator a ball joint, however, it will not help to position the spring.

Answer D is correct. A spring compressor must be used to compress the spring for installation.

Delmar, Cengage Learning ASE Test Preparation

13. The power steering pressure switch is used to:

    A. Increase power steering pressure while driving in a straight line.
    B. Prevent stalling while idling at a stop.
    C. Prevent stalling during a parking maneuver.
    D. Increase power steering pressure while turning.

TASK A.2.7

Answer A is incorrect. The power steering pressure switch does not increase power steering pressure.

Answer B is incorrect. The power steering pressure is low while idling at a stop. The power steering pressure switch signals high pressure.

Answer C is correct. The power steering pressure switch signals the engine control module (ECM) when there is high power steering pressure so the ECM can idle up the engine to prevent stalling.

Answer D is incorrect. The power steering pressure switch does not increase power steering pressure.

14. A vehicle pulls to the right only while accelerating. Which of the following could be the cause?

    A. Underinflated tires
    B. Loose cradle
    C. Loose power steering pump belt
    D. Overinflated tires

TASK B.1.1

Answer A is incorrect. Underinflated tires cause tire wear and sluggish steering response, but will not cause a pull during acceleration.

Answer B is correct. A loose cradle can shift during acceleration and cause a pull.

Answer C is incorrect. A loose power steering pump belt will not cause a pull.

Answer D is incorrect. Overinflated tires will cause tire wear but will not cause a pull during acceleration.

15. A vehicle pulls to the right only while braking. Which of the following could be the cause?

    A. Underinflated tires
    B. Loose cradle
    C. Loose power steering pump belt
    D. Overinflated tires

TASK C.1

Answer A is incorrect. Underinflated tires cause tire wear and sluggish steering response, but will not cause a pull during braking.

Answer B is correct. A loose cradle can shift during braking and cause a pull.

Answer C is incorrect. A loose power steering pump belt will not cause a pull.

Answer D is incorrect. Overinflated tires will cause tire wear but will not cause a pull during braking.

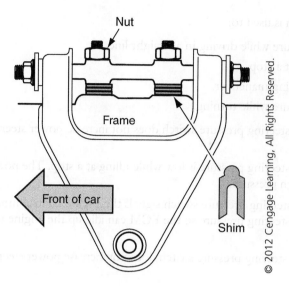

TASK C.6

16. Referring to the figure above, what would be the proper method to add caster without changing camber?

 A. Add shims to both sides.

 B. Remove shims from both sides.

 C. Remove shims from the rear shim pack and add an equal thickness of shims to the front shim pack.

 D. Remove shims from the front shim pack and add an equal thickness of shims to the rear shim pack.

 Answer A is incorrect. This would make camber go negative and not change caster.

 Answer B is incorrect. This would make camber go positive and not affect caster.

 Answer C is incorrect. This would make caster go negative and would not affect camber.

 Answer D is correct. This would make caster go positive and would not affect camber.

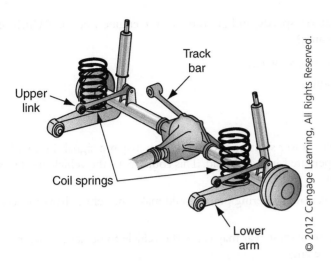

Upper link

Track bar

Coil springs

Lower arm

17. A vehicle equipped with the suspension shown in the figure above pulls to the right on acceleration. Technician A says this could be caused by worn track bar bushings. Technician B says this could be caused by worn bushings in the lower arms. Who is correct?

   A. A only
   B. B only
   C. Both A and B
   D. Neither A nor B

TASK B.2.3

Answer A is incorrect. Technician B is also correct.

Answer B is incorrect. Technician A is also correct.

Answer C is correct. Both Technicians are correct. Both worn track bar and lower arm bushings can allow the axle to shift during acceleration which can cause a pull.

Answer D is incorrect. Both Technicians are correct.

18. There is not enough adjustment left to correct rear camber and rear toe on a vehicle. Which of the following could be the cause?

   A. Sagging rear springs
   B. Worn shock absorbers on the front
   C. Worn shock absorbers on the rear
   D. Sagging front springs

TASK C.12

Answer A is correct. Sagging rear springs will cause ride height to be low which could cause there to be insufficient adjustment room.

Answer B is incorrect. Shock absorbers do not change ride height.

Answer C is incorrect. Shock absorbers do not change ride height.

Answer D is incorrect. Sagging front springs would affect the adjustment on the front.

TASK A.2.7

19. A vehicle has hard steering at low speeds, and occasionally the engine will stall. Which of the following is the most likely cause?

A. Faulty power steering pressure switch

B. Weak power steering pump

C. Worn steering gear

D. Worn ball joints

Answer A is correct. A faulty power steering pressure switch may not signal the ECM to raise the engine idle to compensate for engine load. This could make the vehicle hard to steer and cause the engine to stall.

Answer B is incorrect. A weak power steering pump could make the vehicle hard to steer but would not stall the engine.

Answer C is incorrect. A worn steering gear may cause the vehicle to be hard to steer, however, it would not stall the engine.

Answer D is incorrect. Worn ball joints would not stall the engine.

TASK A.2.8

20. Which of the following would most likely make a vehicle hard to steer?

A. Overtightened power steering pump belt

B. Crimped power steering line

C. Positive camber on the rear

D. Negative caster on the front

Answer A is incorrect. An overtightened power steering pump belt will cause short belt and pump life but will not cause a vehicle to be hard to steer.

Answer B is correct. A crimped power steering line can reduce power steering flow and cause a vehicle to be hard to steer.

Answer C is incorrect. Rear camber will not affect steering effort.

Answer D is incorrect. Negative caster will make a vehicle easier to steer.

TASK A.2.1

21. A vehicle is hard to steer to the left only. Which of the following is the most likely cause?

A. Weak power steering pump

B. Tight steering shaft universal joint

C. Binding ball joint

D. Worn power steering gear

Answer A is incorrect. A weak power steering pump would affect turning in both directions.

Answer B is incorrect. A tight steering shaft universal joint would affect steering feel in both directions.

Answer C is incorrect. A binding ball joint would affect steering feel in both directions.

Answer D is correct. A worn power steering gear could have a leaking internal seal which would allow fluid to bypass in one direction and not in the other. This could make the vehicle hard to steer in one direction only.

22. A vehicle is hard to steer. After disconnecting both outer tie rods, the steering wheel is still hard to turn. Which of the following could be the cause?

TASK A.1.2

   A. Binding strut bearing on the left strut
   B. Worn strut bearing on the right strut
   C. Binding steering column universal joint
   D. Worn rack bushings

   Answer A is incorrect. If the strut bearing was the cause, the steering would be easy after disconnecting the outer tie rod end.

   Answer B is incorrect. If the strut bearing was the cause, the steering would be easy after disconnecting the outer tie rod ends.

   Answer C is correct. Disconnecting the outer tie rod ends would not change a tight, or binding, steering column universal joint.

   Answer D is incorrect. Worn rack bushings usually cause a noise or a looseness in the steering.

23. There is a loud growling noise only when the vehicle is steered to the right while driving. Which of the following could be the cause?

TASK B.1.1

   A. Worn tires
   B. Worn wheel bearing
   C. Incorrect camber
   D. Incorrect toe

   Answer A is incorrect. Worn tires will not cause a loud growling noise when steering to the right.

   Answer B is correct. A loud growling is a typical worn wheel bearing noise.

   Answer C is incorrect. Camber will not cause a noise while turning.

   Answer D is incorrect. Toe will not cause a noise while turning.

24. A vehicle has excessive (6 degrees) negative camber on the left front. The right front camber is within specification. Which of the following could be the cause?

TASK C.3

   A. A worn right rear spring
   B. A worn left rear spring
   C. A worn right front wheel bearing
   D. A worn left front wheel bearing

   Answer A is incorrect. A worn right rear spring would not affect left front camber.

   Answer B is incorrect. A worn left rear spring would not cause the right front camber to be 6 degrees negative.

   Answer C is incorrect. The right front wheel bearing would not affect the left camber.

   Answer D is correct. A worn left front wheel bearing can cause a tire to lean in excessively.

25. A vehicle has 1.5 degrees of cross caster. Technician A says this could cause tire wear. Technician B says this could cause a pull. Who is correct?

TASK C.5

   A. A only
   B. B only
   C. Both A and B
   D. Neither A nor B

   Answer A is incorrect. Technician B is correct. Caster does not cause tire wear.

   Answer B is correct. Only Technician B is correct. Cross caster specifications are usually 0.5 degrees maximum; a cross caster greater than that can cause a pull.

   Answer C is incorrect. Only Technician B is correct.

   Answer D is incorrect. Technician B is correct.

TASK C.3

26. A vehicle has 2.5 degrees cross camber. Technician A says this could cause tire wear. Technician B says this could cause a pull. Who is correct?

    A. A only

    B. B only

    C. Both A and B

    D. Neither A nor B

Answer A is incorrect. Technician B is also correct.

Answer B is incorrect. Technician A is also correct.

Answer C is correct. Both Technicians are correct. Cross camber greater than 0.5 degrees can cause a pull. Also, incorrect camber can cause tire wear.

Answer D is incorrect. Both Technicians are correct.

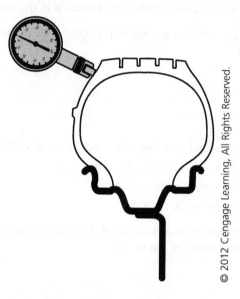

TASK D.6

27. What check is being performed in the figure above?

    A. Lateral runout check

    B. Radial runout check

    C. Static balance check

    D. Dynamic balance check

Answer A is correct. This is a lateral runout check for the wheel and tire combination.

Answer B is incorrect. A radial runout check would need to have the dial indicator positioned on the tread.

Answer C is incorrect. This is not a wheel balance check.

Answer D is incorrect. This is not a wheel balance check.

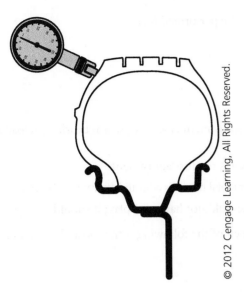

28. Which of the following would be a normal maximum specification for the measurement shown in the figure above?

A. 0.000" (0.00 mm)

B. 0.045" (1.143 mm)

C. 0.090" (2.28 mm)

D. 0.135" (3.43 mm)

TASK D.6

Answer A is incorrect. A normal maximum specification is 0.045" (1.143 mm).

Answer B is correct. A normal maximum specification is 0.045" (1.143 mm).

Answer C is incorrect. A normal maximum specification is 0.045" (1.143 mm).

Answer D is incorrect. A normal maximum specification is 0.045" (1.143 mm).

29. The drive axle nut on a front-wheel drive vehicle is being tightened. Which of the following is true?

A. The vehicle should be on a frame hoist.

B. The nut should be tightened, then backed off 1 flat.

C. The nut should be tightened with an impact wrench.

D. The vehicle should be at normal ride height.

TASK B.1.7

Answer A is incorrect. When possible, all components of the steering and suspension system should be tightened with the vehicle at ride height.

Answer B is incorrect. Normally this nut is torqued to a specification of approximately 80 ft lbs.

Answer C is incorrect. The drive axle can be damaged if the nut is tightened with an impact wrench.

Answer D is correct. When possible, all components of the steering and suspension system should be tightened with the vehicle at ride height.

TASK A.2.11

30. Which steering gear box adjustment should be performed first?

    A. Sector shaft preload

    B. Worm bearing preload

    C. Sector shaft over center

    D. Worm bearing end-play

Answer A is incorrect. Sector shaft preload adjustment is second; worm bearing preload is the first adjustment.

Answer B is correct. Worm bearing preload is the first adjustment.

Answer C is incorrect. Sector shaft over center (total mesh) is adjusted after worm bearing preload.

Answer D is incorrect. The worm bearing should not have end-play; it should have preload.

TASK B.1.1

31. A vehicle sets low on one front corner. Which of the following is the most likely cause?

    A. Weak spring

    B. Weak shock

    C. Bent control arm

    D. Bent steering knuckle

Answer A is correct. Springs control ride height.

Answer B is incorrect. Shocks do not control ride height.

Answer C is incorrect. A bent control arm will affect alignment angles but is not the most likely cause of a vehicle setting low on one front corner.

Answer D incorrect. A bent steering knuckle causes incorrect alignment angle; it does not control ride height.

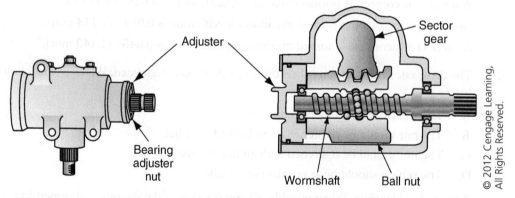

TASK A.2.11

32. Which adjustment is performed with the adjuster shown in the figure above?

    A. Sector shaft preload

    B. Over center adjustment

    C. Worm bearing preload

    D. Sector shaft end-play

Answer A is incorrect. The adjustment shown is not for sector shaft preload. Sector shaft preload is adjusted with the screw on top of the box.

Answer B is incorrect. The over center adjustment is the same as sector shaft preload. It is adjusted with the screw on top of the box.

Answer C is correct. This is a worm bearing preload adjustment.

Answer D is incorrect. The sector shaft should not have end-play.

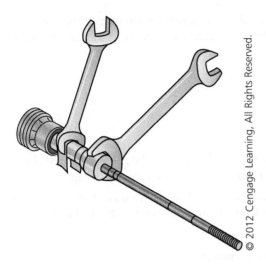

33. What item is being removed in the figure above?

    A. Outer tie rod
    B. Inner tie rod
    C. Pitman arm
    D. Idler arm

TASK A.3.5

Answer A is incorrect. This is the inner tie rod on a rack and pinion steering.

Answer B is correct. This is the inner tie rod on a rack and pinion steering.

Answer C is incorrect. This is not a pitman arm. A pitman arm is used on a recirculating ball steering gear.

Answer D is incorrect. This is not an idler arm. An idler arm is used on a conventional (parallelogram) steering linkage.

34. A customer complains of a vibration only while braking. Technician A says the wheel lug nuts may have been overtorqued. Technician B says the front wheel may be bent. Who is correct?

    A. A only
    B. B only
    C. Both A and B
    D. Neither A nor B

TASK D.4

Answer A is correct. Only Technician A is correct. Overtorqued wheel lug nuts can warp a rotor and cause a vibration while braking.

Answer B is incorrect. Technician A is correct. A bent wheel would not vibrate only while braking.

Answer C is incorrect. Only Technician A is correct.

Answer D is incorrect. Technician A is correct.

35. Which of the following is the most likely cause of tires which are worn in the center of the tread?

    A. Underinflation
    B. Overinflation
    C. Positive caster
    D. Positive camber

TASK D.1

Answer A is incorrect. Underinflation causes wear on the outside shoulders.

Answer B is correct. Overinflation causes wear in the center.

Answer C is incorrect. Caster is not a tire wear angle.

Answer D is incorrect. Positive camber would cause tire wear on the outer shoulders.

TASK A.1.3

36. Technician A says an ohmmeter can be used to measure the resistance of the airbag inflator module. Technician B says a test light should be used to check for voltage at the airbag inflator module. Who is correct?

   A. A only
   B. B only
   C. Both A and B
   D. Neither A nor B

   Answer A is incorrect. Neither Technician is correct. The module should not be checked with an ohmmeter; there is a chance of accidental deployment of the airbag.

   Answer B is incorrect. Neither Technician is correct. The system should not be checked with a test light. This is a computer circuit and a digital multimeter, test light, should be used.

   Answer C is incorrect. Neither Technician is correct.

   Answer D is correct. Neither Technician is correct.

TASK B.1.5

37. A lower ball joint is being inspected on a front suspension. The front coil spring is located between the frame and the lower control arm. Where must the jack stand be placed?

   A. Under the lower control arm
   B. Under the upper control arm
   C. Under the frame
   D. Under the rear cradle

   Answer A is correct. The jack stand must be placed under the lower control arm to unload the ball joint for inspection.

   Answer B is incorrect. The jack stand must be placed under the lower control arm to unload the ball joint for inspection.

   Answer C is incorrect. If the jack stand is placed under the frame, the ball joint would still be loaded.

   Answer D is incorrect. If the jack stand was placed under the rear cradle, the ball joint would still be loaded.

TASK B.2.8

38. A rear tie rod end is being replaced. Which of the following alignment angles will most likely need to be adjusted?

   A. Front toe
   B. Rear toe
   C. Front camber
   D. Front caster

   Answer A is incorrect. A rear tie rod end will not change front toe.

   Answer B is correct. Changing the rear tie rod end will affect rear toe. Rear toe and thrust line will need to be checked and adjusted.

   Answer C is incorrect. A rear tie rod end will not change front camber. Front camber is changed when the front suspension components are replaced.

   Answer D is incorrect. A rear tie rod end will not change front caster. Front caster is changed when the front suspension components are replaced.

39. A rear-wheel drive solid rear axle vehicle has a sheared spring center bolt on the left side. Which of the following will need to be replaced?

TASK B.2.4

 A. The entire left rear spring pack

 B. Both the left and right rear spring packs

 C. The left center bolt

 D. Both left and right center bolts

Answer A is incorrect. The spring pack can have the center bolt replaced.

Answer B is incorrect. The spring pack can have the center bolt replaced.

Answer C is correct. The left spring pack center bolt should be replaced.

Answer D is incorrect. There is no need to replace the right center bolt unless it is damaged.

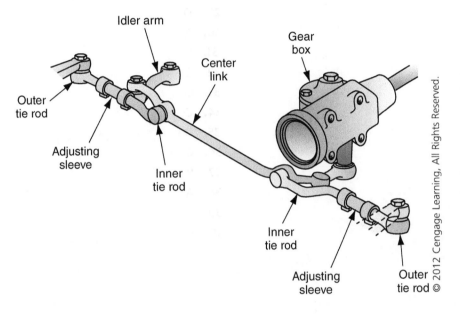

40. The steering linkage illustrated in the figure above is called:

TASK A.2.1

 A. Rack and pinion.

 B. Parallelogram.

 C. Haltenberger.

 D. Relay rod.

Answer A is incorrect. This is a parallelogram (conventional) steering system. A rack and pinion system does not use a recirculating ball gear.

Answer B is correct. This is a parallelogram (conventional) steering system.

Answer C is incorrect. This is a parallelogram (conventional) steering system. A Haltenberger steering system does not have an idler arm.

Answer D is incorrect. This is a parallelogram (conventional) steering system. A relay rod steering system does not have a relay rod.

PREPARATION EXAM 3—ANSWER KEY

1.	B		21.	D
2.	B		22.	A
3.	D		23.	C
4.	C		24.	A
5.	D		25.	B
6.	C		26.	B
7.	A		27.	C
8.	C		28.	C
9.	A		29.	D
10.	D		30.	A
11.	B		31.	C
12.	C		32.	C
13.	B		33.	D
14.	C		34.	C
15.	B		35.	D
16.	A		36.	C
17.	D		37.	B
18.	A		38.	A
19.	D		39.	A
20.	A		40.	D

PREPARATION EXAM 3—EXPLANATIONS

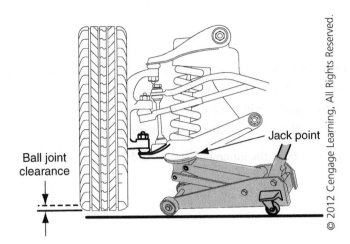

Ball joint clearance

Jack point

© 2012 Cengage Learning, All Rights Reserved.

1. The figure above is measuring:

    A.   The follower ball joint for wear.

    B.   The load carrying ball joint for wear.

    C.   The front coil spring spacer for wear.

    D.   The lower control arm bushings for wear.

    TASK B.1.5

    Answer A is incorrect. The load carrying ball joint is being checked for wear.

    Answer B is correct. The load carrying ball joint is being checked for wear. This is indicated by jacking under the lower control to release the tension on the load carrying joint.

    Answer C is incorrect. The load carrying ball joint is being checked for wear.

    Answer D is incorrect. The lower control arm bushings are checked for wear with the tires on the ground.

2. The left front tire is worn on the inside tread only; the right front tire shows no abnormal wear. Which angle is most likely the cause?

    A.   Caster

    B.   Camber

    C.   Toe-in

    D.   Toe-out

    TASK D.1

    Answer A is incorrect. Caster is not a tire wear angle.

    Answer B is correct. Camber wears the tires on the inner or outer shoulders.

    Answer C is incorrect. Toe-in will usually cause a sawtooth wear pattern.

    Answer D is incorrect. Toe-out will usually cause a sawtooth wear pattern.

TASK D.1

3. The right rear tire has a diagonal swipe wear pattern. This is most likely caused by:

   A. Caster.
   B. Positive camber.
   C. Negative camber.
   D. Toe.

   Answer A is incorrect. Caster is not measured on the rear wheels and is not a tire wear angle.

   Answer B is incorrect. Positive camber will cause outer tread wear.

   Answer C is incorrect. Negative camber will cause inner tread wear.

   Answer D is correct. Rear toe will cause a diagonal swipe on a rear tire.

TASK D.1

4. The left front tire has cupping on both the inner and outer treads. Which of the following is the most likely cause?

   A. Toe-in
   B. Toe-out
   C. Worn shocks
   D. Worn control arm bushings

   Answer A is incorrect. Toe usually causes a sawtooth wear pattern.

   Answer B is incorrect. Toe usually causes a sawtooth wear pattern.

   Answer C is correct. Cupping on both inner and outer treads is usually an indication of worn shocks.

   Answer D is incorrect. Worn control arm bushings will change camber which results in inner tread wear, not cupping.

TASK C.13

5. Thrust line is adjusted by changing:

   A. Rear camber.
   B. Front camber.
   C. Front toe.
   D. Rear toe.

   Answer A is incorrect. Rear camber does not change thrust line.

   Answer B is incorrect. Front camber does not change thrust line.

   Answer C is incorrect. Front toe does not change thrust line. However, front toe should be adjusted to the thrust line.

   Answer D is correct. Adjusting rear toe changes the direction the rear wheels are pointing. Therefore, rear toe changes thrust line.

TASK C.4

6. Technician A says rear camber may be adjusted using shims. Technician B says rear toe may be adjusted using shims. Who is correct?

   A. A only
   B. B only
   C. Both A and B
   D. Neither A nor B

   Answer A is incorrect. Technician B is also correct.

   Answer B is incorrect. Technician A is also correct.

   Answer C is correct. Both Technicians are correct. Rear shims can be used to adjust camber and toe.

   Answer D is incorrect. Both Technicians are correct.

7. Thrust angle is calculated using:

   A. Thrust line and geometric center line.

   B. Camber and SAI.

   C. Thrust line and SAI.

   D. Geometric center line and camber.

TASK C.13

Answer A is correct. Thrust angle is the angle formed by the thrust line and the geometric center line.

Answer B is incorrect. Camber and SAI forms the included angle.

Answer C is incorrect. Thrust line and SAI do not form an angle.

Answer D is incorrect. Geometric center line and camber do not form an angle.

8. Any of these can cause bump steer EXCEPT:

   A. Worn idler arm.

   B. Unlevel steering rack.

   C. Incorrect camber settings.

   D. Worn center link.

TASK A.2.2

Answer A is incorrect. A worn idler arm is a common cause of bump steer.

Answer B is incorrect. An unlevel steering rack is a common cause of bump steer.

Answer C is correct. Camber will cause tire wear and a pull, however, it will not cause bump steer.

Answer D is incorrect. A worn center link will cause bump steer.

9. The tire pressure monitor light on the dash is illuminated. Which of the following is the LEAST LIKELY cause?

   A. Excessive tire pressure in the right rear tire

   B. Low tire pressure in the left front tire

   C. Low tire pressure in the spare

   D. A faulty tire pressure sensor

TASK D.10

Answer A is correct. Most tire pressure monitor systems do not illuminate the light for over-pressurization.

Answer B is incorrect. Low front tire pressure could be the cause.

Answer C is incorrect. Low tire pressure in the spare can be the cause of an illuminated tire pressure monitor system on a five-sensor system.

Answer D is incorrect. A faulty tire pressure sensor can be the cause.

10. Which of the following is true about tire pressure sensors?

   A. The batteries are designed to last a minimum of 15 years.

   B. The batteries are separate replaceable units.

   C. The sensor should be replaced every time the tires are replaced.

   D. The sensors can be part of the valve stem assembly.

TASK D.10

Answer A is incorrect. The batteries are designed to last 10 years.

Answer B is incorrect. The batteries are not replaceable separately from the sensors.

Answer C is incorrect. The sensors do not need to be replaced every time the tires are replaced.

Answer D is correct. The sensors can be part of the valve stem assembly.

TASK A.2.5

11. A customer has a concern about low power assist during parallel park maneuvers. The technician verifies the concern and notes a growling noise coming from the pump. Which of the following is the most likely cause of the concern?

A. Loose power steering belt

B. Low power steering fluid

C. Internally restricted power steering fluid cooler

D. Externally restricted power steering fluid cooler

Answer A is incorrect. A loose power steering belt can cause this concern, however, it would most likely cause a squealing noise, not a growling noise.

Answer B is correct. A low fluid level could cause this condition.

Answer C is incorrect. An internally restricted cooler would cause overheated fluid.

Answer D is incorrect. An externally restricted cooler would cause overheated fluid.

TASK B.2.4

12. A suspension bushing was tightened while the vehicle was supported by a frame hoist. This will cause:

A. Camber wear on the tires.

B. Toe wear on the tires.

C. Short bushing life.

D. The vehicle to pull while driving straight.

Answer A is incorrect. This will not cause the camber to be incorrect.

Answer B is incorrect. This will not cause the toe to be incorrect.

Answer C is correct. This will cause the bushing to be stretched to the extreme when the vehicle is lowered and result in short bushing life.

Answer D is incorrect. This will not cause the vehicle to pull.

TASK B.1.1

13. A front-wheel drive vehicle makes a loud clunk on initial acceleration. Any of these could be the cause EXCEPT:

A. Worn inner CV joint.

B. Worn rack mount bushings.

C. Worn control arm bushings.

D. Loose cradle mounting bolts.

Answer A is incorrect. A worn inner CV joint can cause a loud clunk on acceleration.

Answer B is correct. Worn rack mount bushings can make noise when the steering wheel is turned, but acceleration would not cause them to make noise.

Answer C is incorrect. Worn control arm bushings can cause a loud clunk as the vehicle is accelerated.

Answer D is incorrect. Loose cradle mounting bolts can cause a loud clunk as the vehicle is accelerated.

14. The technician finds the steering wheel moves side to side within the steering column. Which of the following could be the cause?

TASK A.1.2

   A. Worn rack mounts

   B. Worn tie rod ends

   C. Worn upper steering column bearing

   D. Loose steering column mounts

   Answer A is incorrect. Worn rack mounts could cause a noise and a possible loose feel in the steering system but would not cause the wheel to move side to side in the column.

   Answer B is incorrect. Worn tie rod ends will cause play in the steering system but will not cause the steering wheel to move side to side in the steering column.

   Answer C is correct. A worn upper steering column bearing can cause the steering wheel to move side to side within the column.

   Answer D is incorrect. Loose steering column mounts can cause the whole column to move side to side, but not the steering wheel within the column.

15. A vehicle equipped with electric power steering (EPS) has been aligned. Now the vehicle pulls to the left while traveling on a straight, level road. Technician A says this could be caused by an incorrect front toe setting. Technician B says the EPS may need to be re-centered. Who is correct?

TASK A.2.2

   A. A only

   B. B only

   C. Both A and B

   D. Neither A nor B

   Answer A is incorrect. Technician B is correct. Front toe can cause dart, wander, and tire wear; however, it does not cause a pull.

   Answer B is correct. Only Technician B is correct. When a vehicle with EPS is aligned, the system may need to be re-centered using a scan tool. This process allows the ECM to "learn" the straight ahead position.

   Answer C is incorrect. Only Technician B is correct.

   Answer D is incorrect. Technician B is correct.

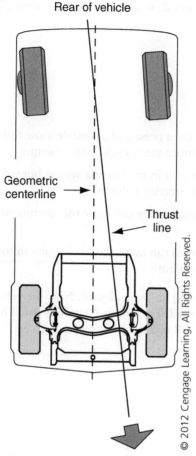

Rear of vehicle

Geometric centerline →

Thrust line ←

TASK C.12

16. What adjustment must be performed to correct the alignment condition shown in the figure above?

   A. Rear toe

   B. Front toe

   C. Rear camber

   D. Front camber

Answer A is correct. Rear toe adjusts the thrust line.

Answer B is incorrect. Front toe does not adjust the thrust line.

Answer C is incorrect. Rear camber adjustment will not correct a thrust line condition.

Answer D is incorrect. Front camber adjustment will not adjust thrust line.

TASK B.2.1

17. While making a hard brake application, the customer notices a clunk from the rear of the vehicle. Which of the following could be the cause?

   A. Leaking struts

   B. A bent rim

   C. An out of round tire

   D. Worn lower control arm bushings

Answer A is incorrect. Leaking struts do not normally clunk.

Answer B is incorrect. A bent rim will cause a vibration, not a clunk.

Answer C is incorrect. An out of round tire will cause a wheel tramp, not a clunk.

Answer D is correct. A worn lower control arm bushing can allow the control arm to move during braking and cause a clunking noise.

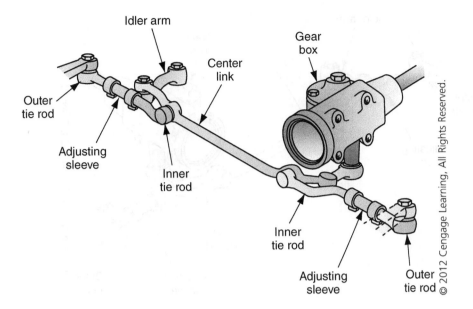

18. A vehicle with the steering/linkage system shown in the figure above needs the left toe adjusted to the positive. Which of the following is true?

    TASK C.7

    A. The adjusting sleeve must be lengthened.

    B. The adjusting sleeve must be shortened.

    C. The steering should be turned 20 degrees to the left during the adjustment.

    D. The steering should be turned 20 degrees to the right during the adjustment.

    Answer A is correct. Since the linkage is behind the front tires, lengthening the adjusting sleeve will create positive, or toe-in.

    Answer B is incorrect. Shortening the sleeve will create negative toe.

    Answer C is incorrect. The steering should be in a straight ahead position when adjusting toe.

    Answer D is incorrect. The steering should be in a straight ahead position when adjusting toe.

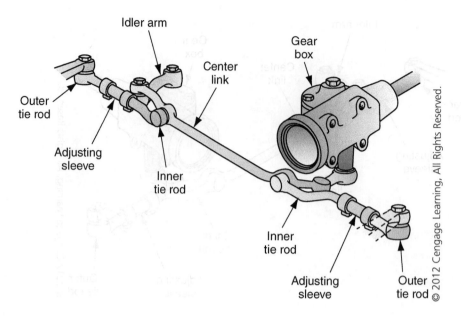

TASK C.7

19. In the figure above, the technician is preparing to turn the adjusting sleeve. Which of the following statements is true?

A. This should be done prior to rear toe.

B. This should be done prior to rear camber.

C. This should be done prior to front camber.

D. This should be done as the last adjustment.

Answer A is incorrect. Rear camber and rear toe are the first two adjustments.

Answer B is incorrect. Rear camber and rear toe are the first toe adjustments.

Answer C is incorrect. Front camber is adjusted before front toe.

Answer D is correct. Front toe is the final alignment adjustment.

TASK C.1

20. A vehicle with the alignment angles shown below is pulling to the left. Which of the following is the most likely cause?

	Spec	Actual		Tolerance
		Left	Right	
Caster	0.2 degrees	0.4 degrees	−1.6 degrees	+/− 0.2
Camber	0.5 degrees	0.6 degrees	0.3 degrees	+/− 0.2
Toe	0.5 degrees	0.3 degrees	0.2 degrees	+/− 0.1

A. Right caster

B. Left caster

C. Left camber

D. Right camber

Answer A is correct. Caster pulls to the least positive. The right front caster is out of specification to the negative.

Answer B is incorrect. Left caster is within tolerance.

Answer C is incorrect. Left camber is within tolerance.

Answer D is incorrect. Right camber is within tolerance.

21.   The vehicle with the alignment readings shown below has a crooked steering wheel. Which of the following is the most likely cause?

TASK C.1

FRONT

| | Spec | Actual | | Tolerance |
		Left	Right	
Caster	1.4 degrees	1.5 degrees	1.3 degrees	+/− 0.2
Camber	0.5 degrees	0.6 degrees	0.4 degrees	+/− 0.2
Toe	0.5 degrees	0.4 degrees	0.4 degrees	+/− 0.1

REAR

| | Spec | Actual | | Tolerance |
		Left	Right	
Camber	0.5 degrees	0.6 degrees	0.3 degrees	+/− 0.2
Toe	0.0 degrees	− 0.3 degrees	0.2 degrees	+/− 0.1

A.   Rear camber

B.   Front camber

C.   Front toe

D.   Rear toe

Answer A is incorrect. Rear camber does not cause a crooked steering wheel.

Answer B is incorrect. Front camber does not cause a crooked steering wheel.

Answer C is incorrect. Front toe can cause a crooked steering wheel, however, this front toe is within specification.

Answer D is correct. Rear toe can cause a crooked steering wheel, and rear toe on this vehicle is out of adjustment.

22.   A vehicle has hard steering at low speeds, and occasionally the engine will stall while parallel parking. Which of the following is the most likely cause?

A.   Dirty throttle body

B.   Weak power steering pump

C.   Worn steering gear

D.   Worn ball joints

TASK A.2.7

Answer A is correct. A dirty throttle body can cause the engine idle speed to be less stable. This could make the vehicle hard to steer and cause the engine to stall.

Answer B is incorrect. A weak power steering pump could make the vehicle hard to steer but would not stall the engine.

Answer C is incorrect. A worn steering gear may cause the vehicle to be hard to steer, however, it would not stall the engine.

Answer D is incorrect. Worn ball joints would not stall the engine.

TASK B.1.1

23. Replacing which of the following will change ride height?

   A. Shocks

   B. Struts

   C. Springs

   D. Sway bars

Answer A is incorrect. Shocks do not change ride height. Shocks help minimize jounce and rebound.

Answer B is incorrect. Struts do not change ride height.

Answer C is correct. Springs affect ride height. Springs support the weight of the vehicle.

Answer D is incorrect. Sway bars do not affect ride height. Sways help to keep the vehicle level during a cornering maneuver.

TASK A.1.3

24. A steering wheel must be removed on a vehicle equipped with a supplemental restraint system (SRS). Technician A says the SRS must have the power removed and the appropriate wait period as prescribed in the service manual must be observed. Technician B says SRS airbag removal can begin as soon as the SRS fuse is removed. Who is correct?

   A. A only

   B. B only

   C. Both A and B

   D. Neither A nor B

Answer A is correct. Only Technician A is correct. The OEM recommended power-down procedures must be followed.

Answer B is incorrect. Technician A is correct. The SRS system has a keep alive power system that must be allowed to dissipate prior to airbag removal.

Answer C is incorrect. Only Technician A is correct.

Answer D is incorrect. Technician A is correct.

TASK A.1.3

25. When storing an airbag on a workbench, which of the following is the correct way to place it?

   A. Face down

   B. Face up

   C. Facing the wall

   D. Facing away from the wall

Answer A is incorrect. Laying it face down will cause the airbag to be propelled in case of accidental deployment.

Answer B is correct. Laying the airbag face up is the safest position.

Answer C is incorrect. Laying the airbag facing the wall will result in the airbag being propelled in case of accidental deployment.

Answer D is incorrect. Laying the airbag facing away from the wall may result in the airbag being propelled in case of accidental deployment.

26. The power steering pressure is being measured. The fluid is warm, the engine is idling, the tester valve is open, and the steering wheel is in the straight ahead position. Which of the following would be considered an acceptable reading?

TASK A.2.7

    A. 0 psi

    B. 100 psi

    C. 500 psi

    D. 1,000 psi

Answer A is incorrect. 0 psi would indicate no pressure. The pressure should be between 50 and 150 psi.

Answer B is correct. Pressure under these conditions should normally be more than 50 and less than 150 psi.

Answer C is incorrect. 500 psi would indicate a restriction in the system.

Answer D is incorrect. 1,000 psi would indicate a restriction in the system.

27. The center link is unlevel on a vehicle with parallelogram steering. Which of the following could be the cause?

TASK A.3.1

    A. Incorrect idler arm height adjustment

    B. Incorrect worm bearing adjust

    C. Incorrect sector shaft adjustment

    D. Incorrect caster adjustment

Answer A is correct. Some idler arms can be adjusted up and down to obtain the correct height. If this adjustment is incorrect, the center link would be unlevel.

Answer B is incorrect. The worm bearing adjustment would not affect the center link being unlevel. It can make the steering loose or tight.

Answer C is incorrect. The sector shaft adjustment would not affect the center link being unlevel. It can cause looseness or binding in the steering.

Answer D is incorrect. Caster would not affect the center link being unlevel. Caster can cause a pull.

28. Which of the following would be used to remove a pitman arm?

TASK A.3.2

    A. Pickle fork

    B. Ball joint separator

    C. Puller

    D. Slide hammer

Answer A is incorrect. A pickle fork is usually used to separate ball joints or tie rod ends.

Answer B is incorrect. A ball joint separator is used to separate ball joints.

Answer C is correct. A puller specially designed to remove the pitman arm is used.

Answer D is incorrect. A slide hammer is not used to remove the pitman arm.

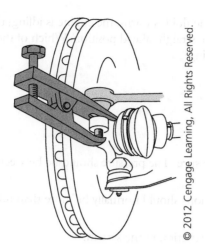

TASK A.3.5

29. What is being performed in the figure above?

    A. Idler arm removal
    B. Pitman arm removal
    C. Ball joint removal
    D. Tie rod end removal

    Answer A is incorrect. The idler arm is connected to the frame and the center link.

    Answer B is incorrect. The pitman arm is connected to the steering gear and the center link.

    Answer C is incorrect. The ball joint is connected to the control arm and the steering knuckle.

    Answer D is correct. The tie rod end is being removed.

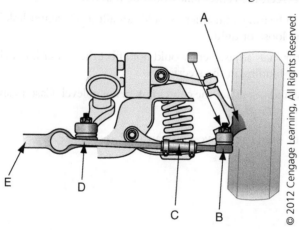

TASK C.7

30. Item D in the figure above was bent during an accident. Which of the following would occur?

    A. Toe would go positive.
    B. Toe would go negative.
    C. Caster would go positive.
    D. Caster would go negative.

    Answer A is correct. If item D was bent, the tie rod assembly would be shortened which would cause toe-in (positive toe) on this vehicle.

    Answer B is incorrect. If item D was bent, the tie rod assembly would be shortened which would cause toe-in (positive toe), not toe-out (negative) on this vehicle.

    Answer C is incorrect. The tie rod assembly does not change caster.

    Answer D is incorrect. The tie rod assembly does not change caster.

31.　Both front tires on a solid front axle vehicle have excessive positive camber. Which of the following could be the cause?

TASK B.1.6

　　A.　Caster shims were installed.

　　B.　The toe was improperly adjusted.

　　C.　The axle is bent.

　　D.　The drag link is bent.

Answer A is incorrect. Caster shims would affect caster, not camber.

Answer B is incorrect. Toe adjustment does not affect camber.

Answer C is correct. A bent front axle could cause both front tires to have excessive camber.

Answer D is incorrect. A bent drag link would change the toe, not the camber.

32.　The left front leaf spring is being replaced on a solid front axle vehicle. Which of the following should also be changed?

TASK B.1.9

　　A.　The left front shock

　　B.　Both front shocks

　　C.　The right front leaf spring

　　D.　The left rear spring

Answer A is incorrect. The shock does not necessarily need to be replaced just because the spring is replaced.

Answer B is incorrect. The front shocks do not have to be replaced because of a spring replacement.

Answer C is correct. The front leaf springs must be replaced in pairs.

Answer D is incorrect. The left rear leaf spring does not have to be replaced because the left front leaf spring was replaced.

33.　A customer is concerned that the rear suspension sags when there are passengers in the rear seats. Which of the following is the most likely cause?

TASK B.2.1

　　A.　Weak shocks

　　B.　Worn sway bar

　　C.　Worn jounce bumpers

　　D.　Weak springs

Answer A is incorrect. Shocks do not control ride height; they control jounce and rebound.

Answer B is incorrect. Sway bars do not control ride height; they help control body roll during a turn.

Answer C is incorrect. Jounce bumpers do not control ride height; they limit downward travel.

Answer D is correct. Springs control ride height. Weak springs could allow the vehicle to set noticeably low when a load is in the rear of the vehicle.

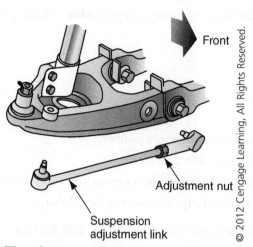

Front

Adjustment nut

Suspension
adjustment link

TASK B.2.8

34. The adjustment link shown in the figure above is worn. This would most likely result in:

   A.   A change in front toe.
   B.   A change in front camber.
   C.   A change in rear toe.
   D.   A change in rear camber.

   Answer A is incorrect. This is a rear suspension system.

   Answer B is incorrect. This is a toe adjuster on the rear.

   Answer C is correct. This is the rear toe adjuster.

   Answer D is incorrect. This is not the rear camber adjustment.

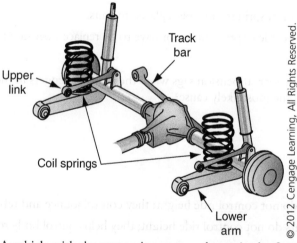

Track
bar

Upper
link

Coil springs

Lower
arm

TASK B.2.3

35. A vehicle with the suspension system shown in the figure above has excessive driveline vibrations only when loaded. Which of the following could be the cause?

   A.   Worn track bar bushings
   B.   A bent track bar
   C.   A bent lower control arm
   D.   Worn upper link bushings

   Answer A is incorrect. Worn track bar bushings would cause the vehicle to lose directional stability during cornering

   Answer B is incorrect. A bent track bar would cause the rear axle to be offset.

   Answer C is incorrect. A bent lower control arm would change the thrust line.

   Answer D is correct. Worn upper link bushings could cause the driveline pinion angle to change under load, which would result in a vibration.

36. A solid rear axle vehicle has a shorter wheelbase on one side than the other. Technician A says this could cause a pull to the side with the shorter wheelbase. Technician B says this could cause a thrust angle problem. Who is correct?

    TASK B.2.1

    A. A only
    B. B only
    C. Both A and B
    D. Neither A nor B

    Answer A is incorrect. Technician B is also correct.

    Answer B is incorrect. Technician A is also correct.

    Answer C is correct. Both Technicians are correct. Setback can cause the vehicle to pull to the side with the shorter wheelbase. Also, when there is a wheelbase difference on a solid rear axle vehicle, the rear axle can be shifted, resulting in a thrust angle condition.

    Answer D is incorrect. Both Technicians are correct.

37. Which of the following would be the most normal end-play specification for a set of tapered roller front-wheel bearings?

    TASK C.2

    A. 0.0001″–0.0005″ (0.00254 mm–0.0127 mm)
    B. 0.0010″–0.0050″ (0.0254 mm–0.127 mm)
    C. 0.0100″–0.0500″ (0.254 mm–1.27 mm)
    D. 0.1000″–0.5000″ (2.54 mm–12.7 mm)

    Answer A is incorrect. This would be too little clearance.

    Answer B is correct. This is the correct specification.

    Answer C is incorrect. This would be too much clearance.

    Answer D is incorrect. This would be too much clearance.

38. Which of the following would be the most normal cross caster specification?

    TASK C.5

    A. No more than 0.5 degrees
    B. No less than 0.5 degrees
    C. No more than 1.5 degrees
    D. No less than 1.5 degrees

    Answer A is correct. The most normal cross caster specification is no more 0.5 degrees.

    Answer B is incorrect. The most normal cross caster specification is no more 0.5 degrees.

    Answer C is incorrect. The most normal cross caster specification is no more 0.5 degrees.

    Answer D is incorrect. The most normal cross caster specification is no more 0.5 degrees.

39. Which of the following would be the most normal cross camber specification?

    TASKS C.3, B.1.2

    A. No more than 0.5 degrees
    B. No less than 0.5 degrees
    C. No more than 1.5 degrees
    D. No less than 1.5 degrees

    Answer A is correct. The most normal cross camber specification is no more 0.5 degrees.

    Answer B is incorrect. The most normal cross camber specification is no more 0.5 degrees.

    Answer C is incorrect. The most normal cross camber specification is no more 0.5 degrees.

    Answer D is incorrect. The most normal cross camber specification is no more 0.5 degrees.

TASK C.5

40. Technician A says the vehicle with the specifications below will pull to the left due to the caster settings. Technician B says the vehicle will wear tires due to the caster settings. Who is correct?

FRONT

	Spec	Actual		Tolerance
		Left	Right	
Caster	2.4 degrees	1.5 degrees	1.3 degrees	+/− 0.5
Camber	0.5 degrees	0.5 degrees	0.4 degrees	+/− 0.2
Toe	0.5 degrees	0.25 degrees	0.25 degrees	+/− 0.2

REAR

	Spec	Actual		Tolerance
		Left	Right	
Camber	0.5 degrees	0.6 degrees	0.5 degrees	+/− 0.2
Toe	0.0 degrees	− 0.1 degrees	0.1 degrees	+/− 0.1

A. A only

B. B only

C. Both A and B

D. Neither A nor B

Answer A is incorrect. Neither Technician is correct. Camber settings are below specification, but cross camber is less than 0.5 degrees. This vehicle will not pull due to caster.

Answer B is incorrect. Neither Technician is correct. Caster is not a tire wear setting.

Answer C is incorrect. Neither Technician is correct.

Answer D is correct. Neither Technician is correct.

PREPARATION EXAM 4—ANSWER KEY

1.	B	21.	B
2.	B	22.	D
3.	C	23.	A
4.	D	24.	B
5.	A	25.	B
6.	B	26.	C
7.	C	27.	D
8.	C	28.	D
9.	D	29.	D
10.	A	30.	C
11.	C	31.	C
12.	C	32.	B
13.	D	33.	A
14.	A	34.	D
15.	C	35.	C
16.	A	36.	B
17.	C	37.	C
18.	B	38.	B
19.	D	39.	C
20.	C	40.	D

PREPARATION EXAM 4–EXPLANATIONS

1. The tire pressure monitor system (TPMS) light is illuminated on the dash. Which of the following could be the cause?

TASK D.10

   A. The traction control system has a code.

   B. One tire has low tire pressure.

   C. The torsion bars sense an overload.

   D. The tires are due to be rotated.

   Answer A is incorrect. The TPMS light indicates low tire pressure has been measured or there is a problem in the tire pressure monitoring system. The traction control system does not illuminate the TPMS light.

   Answer B is correct. The TPMS light indicates low tire pressure has been measured or there is a problem in the tire pressure monitoring system.

   Answer C is incorrect. The TPMS light indicates low tire pressure has been measured or there is a problem in the tire pressure monitoring system. The torsion bars do not activate the TPMS light.

   Answer D is incorrect. The TPMS light indicates low tire pressure has been measured or there is a problem in the tire pressure monitoring system. The TPMS light is not a maintenance reminder light.

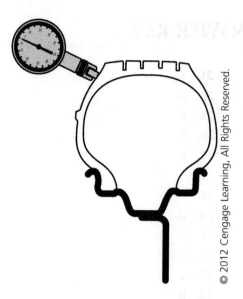

TASK D.6

2. The measurement shown in the figure above is out of specification. Technician A says an out of round brake drum could be the cause. Technician B says a bent rim could be the cause. Who is correct?

    A. A only

    B. B only

    C. Both A and B

    D. Neither A nor B

Answer A is incorrect. Technician B is correct. An out of round brake drum will cause brake pedal pulsation while braking; it will not cause the wheel/tire lateral runout to be excessive.

Answer B is correct. Only Technician B is correct. A bent rim can cause the lateral runout to be excessive.

Answer C is incorrect. Only Technician B is correct.

Answer D is incorrect. Technician B is correct.

TASK D.4

3. An out of round rear tire would cause what type of customer concern?

    A. A vibration that tends to gets worse at higher road speeds

    B. A vibration in the steering wheel

    C. A vibration which is worse at low speeds and tends to get better at high speeds

    D. A vibration that only occurs during cornering

Answer A is incorrect. An out of round tire vibration is more noticeable at low speeds.

Answer B is incorrect. A vibration in the steering wheel is usually caused by front tires.

Answer C is correct. An out of round tire vibration is less noticeable at high speeds because the low spot on the tire tends to be skipped over.

Answer D is incorrect. Vibrations during cornering are usually bearing or CV joint concerns.

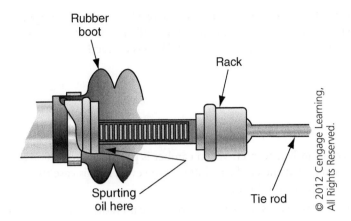

Rubber boot

Rack

Spurting oil here

Tie rod

© 2012 Cengage Learning, All Rights Reserved.

4. Fluid is spurting as shown in the figure above. Which of the following is the most likely repair procedure?

    A.   Replace the tie rod.
    B.   Replace the rubber boot.
    C.   Replace the inner rack seal.
    D.   Replace the rack.

TASK A.2.2

Answer A is incorrect. Replacing the tie rod would not repair the leak.

Answer B is incorrect. Replacing the boot would not repair the leak.

Answer C is incorrect. Replacing the seal may repair the leak, however, it is very rare for a technician to replace inner rack seals.

Answer D is correct. The most normal repair is to replace the rack.

5. Which of the following would be the most likely cause of a steering pull?

    A.   Worn control arm bushings
    B.   Broken sway bar end links
    C.   Worn shock absorber bushings
    D.   Incorrect front toe settings

TASK C.1

Answer A is correct. Worn control arm bushings will allow changes to camber and caster which can cause the vehicle to pull.

Answer B is incorrect. The sway bar end links will cause an occasional knocking noise but will not cause a vehicle to pull.

Answer C is incorrect. Worn shock absorber bushings will cause noise but will not affect steering pull.

Answer D is incorrect. Incorrect front toe settings can cause tie wear, and perhaps a wander or darting condition, but will not cause a pull.

6. A vehicle alignment is checked. The left rear camber is excessively positive. The right rear camber is excessively negative. Technician A says the front cradle may need to be shifted. Technician B says the rear cradle may need to be shifted. Who is correct?

    A. A only

    B. B only

    C. Both A and B

    D. Neither A nor B

Answer A is incorrect. Technician B is correct. When camber is positive on one side and negative on the other, it is likely that the cradle is shifted. However, in this case the alignment concern is in the rear, not the front.

Answer B is correct. Only Technician B is correct. A shifted rear cradle could cause this condition.

Answer C is incorrect. Only Technician B is correct.

Answer D is incorrect. Technician B is correct.

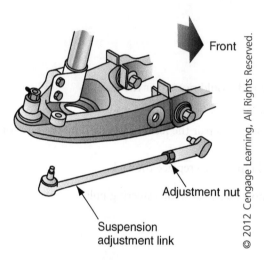

Front

Adjustment nut

Suspension
adjustment link

© 2012 Cengage Learning, All Rights Reserved.

7. The adjustment link shown in the figure above is used to adjust:

    A. Front camber.

    B. Front toe.

    C. Rear toe.

    D. Rear camber.

Answer A is incorrect. This is the rear of the vehicle.

Answer B is incorrect. This is the rear of the vehicle.

Answer C is correct. This is a rear toe adjuster.

Answer D is incorrect. This is a rear toe adjuster, not rear camber.

8. Which of the following would be the most likely maximum radial runout specification on a wheel tire assembly for a passenger vehicle?

   A. 0.020"

   B. 0.030"

   C. 0.050"

   D. 0.090"

TASK D.6

Answer A is incorrect. 0.020" would not be enough radial runout to cause a problem.

Answer B is incorrect. 0.030" would not be enough radial runout to cause a problem.

Answer C is correct. 0.050" could be felt on some vehicles, especially on the front of a vehicle. This is most likely maximum radial runout specification.

Answer D is incorrect. 0.090" would be too much radial runout on a passenger vehicle.

9. A technician is preparing to measure radial runout on a wheel tire assembly. Where should the dial indicator be placed?

   A. On the rim bead seat parallel to the axle

   B. On the rim bead seat perpendicular to the axle

   C. On the tire parallel to the axle

   D. On the tire perpendicular to the axle

TASK D.6

Answer A is incorrect. This placement would measure rim radial runout.

Answer B is incorrect. This placement would measure rim lateral runout.

Answer C is incorrect. This placement would measure tire lateral runout.

Answer D is correct. This placement would measure tire and wheel radial runout.

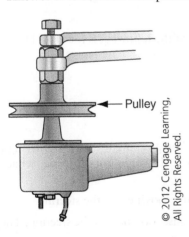

Pulley

© 2012 Cengage Learning, All Rights Reserved.

10. Which procedure is being performed in the figure above?

    A. Replacing a power steering pump pulley

    B. Adjusting power steering pump belt tension

    C. Mounting the power steering pump

    D. Preparing the pump for pressure testing

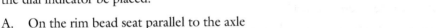

TASK A.2.4

Answer A is correct. The power steering pump pulley is being replaced.

Answer B is incorrect. This is not how the power steering pump belt tension is adjusted.

Answer C is incorrect. The power steering pump is not being mounted.

Answer D is incorrect. The pulley does not need to be changed to measure pump pressure.

TASK C.2

11. Tapered roller wheel bearings are being serviced. Which of the following is true?

　　A. The cup is always replaced.

　　B. The hub must be held still while the bearing is adjusted.

　　C. The seal is always replaced.

　　D. The bearing should be adjusted with the tire on the ground.

Answer A is incorrect. The cup is replaced only if damaged.

Answer B is incorrect. The hub is turned during the adjustment process.

Answer C is correct. The seal is always replaced.

Answer D is incorrect. The bearing should be adjusted while spinning the hub.

TASK A.3.5

12. A tie rod end has been installed and the nut properly torqued. The cotter pin hole is not aligned. Which of the following should the technician do?

　　A. Replace the tie rod end.

　　B. Replace the steering knuckle.

　　C. Tighten the nut until the hole aligns.

　　D. Loosen the nut until the hole aligns.

Answer A is incorrect. The nut should be tightened until the hole aligns. There is no need to replace the tie rod end.

Answer B is incorrect. The nut should be tightened until the hole aligns. There is no need to replace the steering knuckle.

Answer C is correct. The nut should be tightened until the hole aligns.

Answer D is incorrect. The nut should be tightened until the hole aligns. Loosening the nut will allow clearance between the stud and hole.

TASK A.2.10

13. There is excessive play in the steering system. Which of the following is the most likely cause?

　　A. Worn strut mounts

　　B. Worn sway bar bushings

　　C. Worn shock mounts

　　D. Worn rack mounts

Answer A is incorrect. Worn strut mounts are located by grasping the strut and trying to move it, or listening for noise while steering the vehicle. It would not cause play while turning the steering wheel.

Answer B is incorrect. Sway bar bushings will not cause looseness in the steering.

Answer C is incorrect. Worn shock mounts will not cause looseness in the steering. They can cause a knocking sound while in motion.

Answer D is correct. Worn rack mounts will allow the rack to shift which will cause a looseness in the steering.

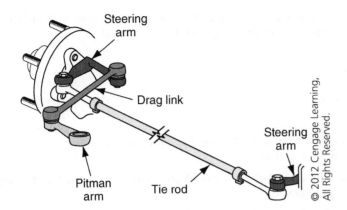

14. Referring to the figure above, which of the following would be used to adjust total toe?

    A. Tie rod
    B. Steering arm
    C. Drag link
    D. Pitman arm

TASK C.7

Answer A is correct. Total toe is adjusted by turning the tie rod.

Answer B is incorrect. Total toe is adjusted by turning the tie rod. Steering arms are not adjustable.

Answer C is incorrect. The drag link is adjusted to center the steering wheel.

Answer D is incorrect. Total toe is adjusted by turning the tie rod. Pitman arms are not adjustable.

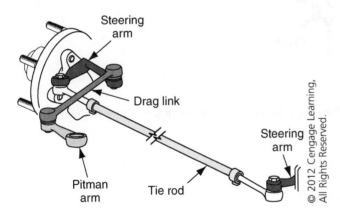

15. Referring to the figure above, which of the following would be used to center the steering wheel?

    A. Tie rod
    B. Steering arm
    C. Drag link
    D. Pitman arm

TASK C.8

Answer A is incorrect. Total toe is adjusted by turning the tie rod.

Answer B is incorrect. The steering arm is not adjustable.

Answer C is correct. The drag link is adjusted to center the steering wheel.

Answer D is incorrect. The pitman arm is not adjustable.

16. The steering binds while turning in either direction. The pitman arm is removed and the steering continues to bind. Which of the following could be the cause?

    A. A binding steering shaft U-joint
    B. A binding upper ball joint
    C. A binding lower ball joint
    D. A binding tie rod end

    Answer A is correct. A binding steering shaft U-joint could continue to cause the steering to bind after the pitman arm was removed.

    Answer B is incorrect. A binding upper ball joint would not continue to cause the steering to bind after the pitman arm was removed.

    Answer C is incorrect. A binding lower ball joint would not continue to cause the steering to bind after the pitman arm was removed.

    Answer D is incorrect. A binding tie rod would not continue to cause the steering to bind after the pitman arm was removed.

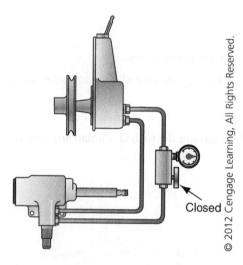

Closed

17. What is being measured when the valve illustrated in the figure above is closed?

    A. Fluid flow (GPH)
    B. Pump vacuum (Hg)
    C. Pump pressure (PSI)
    D. Fluid temperature (F)

    Answer A is incorrect. Fluid flow is measured with the valve open.

    Answer B is incorrect. The pump vacuum is not checked on a power steering pump.

    Answer C is correct. Maximum pump pressure is being checked.

    Answer D is incorrect. Fluid temperature is checked with a thermometer in the reservoir.

18. The steering knuckle is being replaced on the front of a vehicle with torsion bar SLA suspension. Any of these is true concerning this procedure EXCEPT:

    A. The outer tie rod must be disconnected.

    B. The inner tie rod must be disconnected.

    C. The upper ball joint must be disconnected.

    D. The lower ball joint must be disconnected.

TASK B.1.7

Answer A is incorrect. The outer tire rod will need to be disconnected to replace the steering knuckle.

Answer B is correct. It will not be necessary to remove the inner tie rod to replace the steering knuckle.

Answer C is incorrect. It will be necessary to disconnect the upper ball joint to remove the steering knuckle.

Answer D is incorrect. It will be necessary to disconnect the lower ball joint to remove the steering knuckle.

19. Any of these is true about ball joint replacement EXCEPT:

    A. The ball joint may be pressed in.

    B. The ball joint may be bolted in.

    C. The control arm may have to be replaced with the ball joint.

    D. The inner tie rod may have to be replaced with the ball joint.

TASK B.1.5

Answer A is incorrect. The ball joint may be pressed in.

Answer B is incorrect. The ball joint may be bolted in.

Answer C is incorrect. Some ball joints are not separately replaceable from the control arm. In these models, the control arm must be replaced to replace the ball joint.

Answer D is correct. The inner tie rod is not part of a ball joint replacement.

20. The sector shaft adjustment is being performed on a recirculating ball steering gear. Technician A says turning torque should be measured with an inch/pound torque wrench. Technician B says the adjustment is changed by rotating the adjusting screw on the top of the steering gear. Who is correct?

TASK A.2.11

    A. A only

    B. B only

    C. Both A and B

    D. Neither A nor B

Answer A is incorrect. Technician B is also correct.

Answer B is incorrect. Technician A is also correct.

Answer C is correct. Both Technicians are correct. The torque is checked using an inch/pound torque wrench and the adjusting screw is located on the top of the steering gear.

Answer D is incorrect. Both Technicians are correct.

TASKS A.3.4, C.7

21. An idler arm is being replaced. Which of the following alignment angles will need to be checked after replacement?

    A. Caster

    B. Toe

    C. Camber

    D. Thrust line

Answer A is incorrect. An idler arm will not change caster

Answer B is correct. The idler arm supports the center link, therefore changing an idler arm will change toe.

Answer C is incorrect. Camber will not be affected.

Answer D is incorrect. Thrust line is established by rear toe. Thrust line will not be affected.

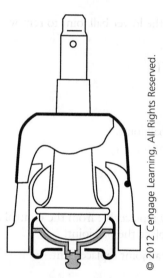

TASK B.1.5

22. The ball joint shown in the figure above is being inspected. Technician A says the vehicle will need to be lifted on a frame hoist. Technician B says the ball joint must be removed from the vehicle to be properly inspected. Who is correct?

    A. A only

    B. B only

    C. Both A and B

    D. Neither A nor B

Answer A is incorrect. Neither Technician is correct. The vehicle must be setting on the tires.

Answer B is incorrect. Neither Technician is correct. This type of ball joint must be in the vehicle to be inspected.

Answer C is incorrect. Neither Technician is correct.

Answer D is correct. Neither Technician is correct. This is a wear indicator ball joint and it should be inspected while still in the vehicle with the vehicle setting on the tires.

23. A vehicle becomes unpredictable when cornering. Which of the following is the most likely cause?

    A. Worn track bar bushings
    B. Worn shock bushings
    C. Weak front springs
    D. Weak rear springs

    TASK B.1.4

    Answer A is correct. Worn track bar bushings will allow the axle to shift during cornering which will change axle offset and make the vehicle unstable.

    Answer B is incorrect. Worn shock bushings will cause noise and cause the vehicle to bounce.

    Answer C is incorrect. Weak front springs will affect ride height.

    Answer D is incorrect. Weak rear springs will affect ride height.

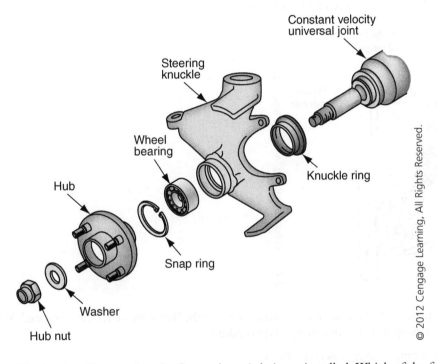

© 2012 Cengage Learning, All Rights Reserved.

24. The bearing illustrated in the figure above is being reinstalled. Which of the following is correct concerning installing the hub nut?

    A. The hub nut can be reused.
    B. The hub nut should be torqued with the wheels on the ground.
    C. The hub nut should be torqued, then backed off one-half turn.
    D. Locking compound should be used on the hub nut.

    TASK C.2

    Answer A is incorrect. While some hub nuts use a lock with a cotter pin, this one does not. Therefore, this hub nut is a self-locking nut and should not be reused.

    Answer B is correct. The hub nut should be torqued with the wheels on the ground when possible.

    Answer C is incorrect. The hub nut is not backed off after torquing.

    Answer D is incorrect. This is a self-locking nut; locking compound is not necessary.

TASK B.1.1

25. Technician A says ride height must be measured both before and after an alignment. Technician B says ride height should be measured before an alignment. Who is correct?

A. A only

B. B only

C. Both A and B

D. Neither A nor B

Answer A is incorrect. Technician B is correct. Ride height will not change during an alignment.

Answer B is correct. Only Technician B is correct. Ride height is measured during a pre-alignment inspection. If ride height is low, corrections must be made prior to the alignment.

Answer C is incorrect. Only Technician B is correct.

Answer D is incorrect. Technician B is correct.

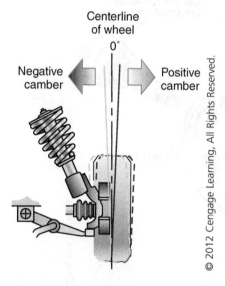

Centerline
of wheel
0°

Negative camber

Positive camber

TASK B.1.8

26. The spring in the front suspension system shown in the figure above is being replaced. Which of the following is true concerning this procedure?

A. The spring is compressed on the vehicle.

B. The strut must also be replaced.

C. The alignment will need to be checked after spring replacement.

D. The rear spring on the same side must also be replaced.

Answer A is incorrect. The spring is compressed after the strut assembly is removed from the vehicle.

Answer B is incorrect. The strut can be replaced separately from the spring.

Answer C is correct. The alignment will be changed by the replacement of the spring and therefore must be checked.

Answer D is incorrect. The rear spring does not have to be replaced.

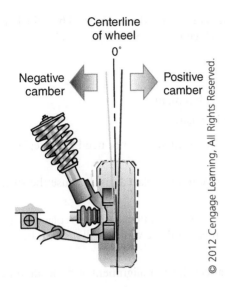

Centerline
of wheel
0°

Negative
camber

Positive
camber

27. The spring in the front suspension system shown in the figure above is being replaced. Which of the following is true concerning this procedure?

TASK B.1.8

A. The spring is compressed on the vehicle.

B. The strut must also be replaced.

C. The rear alignment toe angle will be changed.

D. The front spring on the other side must also be replaced.

Answer A is incorrect. The spring is compressed after the strut assembly is removed from the vehicle.

Answer B is incorrect. The strut can be replaced separately from the spring.

Answer C is incorrect. Changing the front coil spring will not change rear toe.

Answer D is correct. The front spring on the other side must also be replaced. Springs are replaced across the axle.

28. A vehicle with coil spring front suspension has a low ride height. Technician A says the springs can be adjusted to correct ride height. Technician B says the springs can be rotated to correct ride height. Who is correct?

TASK B.1.8

A. A only

B. B only

C. Both A and B

D. Neither A nor B

Answer A is incorrect. Coil springs are not adjusted.

Answer B is incorrect. Coil springs are not rotated.

Answer C is incorrect. Neither Technician is correct.

Answer D is correct. Neither Technician is correct.

TASK B.2.6

29. A customer has requested that the rear struts be replaced on their vehicle. The vehicle has McPherson strut independent rear suspension. Which of the following is true?

    A. The strut bearings must also be replaced.

    B. The rear springs must also be replaced.

    C. The alignment must be corrected prior to replacement.

    D. The rear alignment must be checked after installation.

    Answer A is incorrect. The strut bearings are on the front and do not need to be replaced just because the struts are replaced.

    Answer B is incorrect. The rear springs do not have to be replaced just because the struts are replaced.

    Answer C is incorrect. It is not necessary to correct the alignment prior to replacing the struts. Replacing the struts will change the alignment and the vehicle will need to be realigned after the new struts are installed.

    Answer D is correct. Replacing the struts will affect the rear alignment and the alignment must be checked after installation.

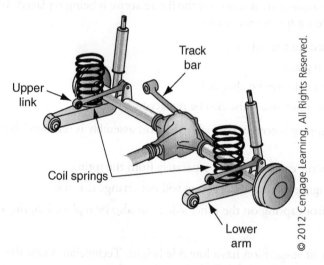

TASK B.2.3

30. The shocks are being replaced on the suspension system shown in the figure above. Which of the following is true regarding this procedure?

    A. The upper links will need to be disconnected.

    B. The lower arms will need to be disconnected.

    C. If a frame hoist is used, the axle will need to be supported.

    D. The track bar must also be replaced.

    Answer A is incorrect. The upper links do not need to be disconnected to replace the shocks.

    Answer B is incorrect. The lower arms do not need to be disconnected. They will not interfere with shock replacement.

    Answer C is correct. Since the shocks provide the stops for the full extension of this suspension, the axle will need to be supported.

    Answer D is incorrect. The track bar does not need to be replaced just because the shocks are replaced.

31. A jounce bumper on the rear suspension is worn excessively. Which of the following is the most likely cause?

    A. Incorrect toe settings
    B. Incorrect camber settings
    C. Worn springs
    D. Worn strut rod bushings

TASK B.2.5

    Answer A is incorrect. Alignment angles (toe) do not affect jounce bumpers. Jounce bumpers are worn by suspension travel.

    Answer B is incorrect. Alignment angles (camber) do not affect jounce bumpers. Jounce bumpers are worn by suspension travel.

    Answer C is correct. Worn springs will allow the vehicle suspension to settle and the suspension to contact the jounce bumpers; this can cause worn jounce bumpers.

    Answer D is incorrect. Worn strut rod bushings will not affect the jounce bumpers; they can affect alignment angles.

32. An alignment is being performed on a McPherson strut vehicle. Technician A says the toe adjustment may be at the base of the strut. Technician B says the camber adjustment may be at the strut mount. Who is correct?

    A. A only
    B. B only
    C. Both A and B
    D. Neither A nor B

TASK C.4

    Answer A is incorrect. Technician B is correct. The toe adjustment is performed at the tie rod.

    Answer B is correct. Only Technician B is correct. Some camber adjustments are performed at the strut mount.

    Answer C is incorrect. Only Technician B is correct.

    Answer D is incorrect. Technician B is correct.

33. A vehicle has a negative setback. Which of the following is true?

    A. The vehicle may pull to the left.
    B. The vehicle may pull to the right.
    C. The left rear tire will have a diagonal swipe wear pattern.
    D. The right rear tire will have an outside tread wear pattern.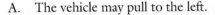

TASK C.14

    Answer A is correct. The left wheel is further back than the right in a negative setback condition. This means the left wheelbase is shorter than the right. A vehicle can pull to the side with the shortest wheelbase.

    Answer B is incorrect. If there is a pull caused by negative setback, it will be to the left.

    Answer C is incorrect. A diagonal swipe wear pattern is caused by rear toe.

    Answer D is incorrect. An outside tread wear pattern is caused by positive camber.

TASK C.10

34. Steering axis inclination is out of specification. Which of the following could be the cause?

   A. A bent steering arm
   B. Incorrect camber settings
   C. Incorrect caster settings
   D. A bent strut

Answer A is incorrect. A bent steering arm can cause toe-out-on-turns to be incorrect.

Answer B is incorrect. Camber settings do not change SAI.

Answer C is incorrect. Caster settings do not change SAI.

Answer D is correct. A bent strut will change the SAI readings and can cause SAI to be out of specification.

TASK C.3

35. Camber is found to be out of adjustment on the left front of a vehicle. There is not enough adjustment remaining to correct the alignment. Any of these could be the cause EXCEPT:

   A. A bent control arm.
   B. Incorrect ride height.
   C. Incorrect thrust angle.
   D. A bent spindle.

Answer A is incorrect. A bent control arm can affect the camber angle.

Answer B is incorrect. Incorrect ride height can affect camber.

Answer C is correct. Thrust angle does not change camber.

Answer D is incorrect. A bent spindle can affect camber.

TASK C.12

36. Excessive rear positive toe is found on both sides of a front-wheel drive vehicle with a solid rear axle. Which of the following could be the cause?

   A. A bent front axle
   B. A bent rear axle
   C. A bent lower control arm on the rear
   D. A bent upper control arm on the rear

Answer A is incorrect. A bent front axle would change the front alignment angles, not the rear.

Answer B is correct. A bent rear axle can cause the toe to be positive on both rear tires.

Answer C is incorrect. A front-wheel drive with a solid rear axle vehicle does not have lower control arms on the rear.

Answer D is incorrect. A front-wheel drive with a solid rear axle vehicle does not have upper control arms on the rear.

37. During an alignment, the technician finds caster to be below specification. Technician A says this could be caused by an incorrect rear ride height setting. Technician B says this could be caused by a malfunctioning air suspension system. Who is correct?

TASK C.5

   A.   A only
   B.   B only
   C.   Both A and B
   D.   Neither A nor B

   Answer A is incorrect. Technician B is also correct.

   Answer B is incorrect. Technician A is also correct.

   Answer C is correct. Both Technicians are correct. Higher than specification rear ride height would cause caster to go negative. Higher than normal rear ride height can be caused by an air suspension system that will not deflate the airbags when appropriate.

   Answer D is incorrect. Both Technicians are correct.

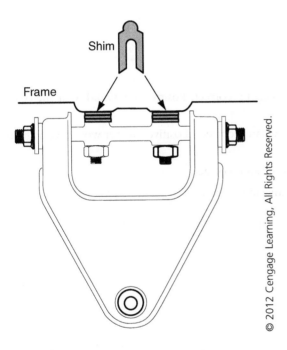

38. Camber needs to be moved toward the positive on the suspension system shown in the figure above. Caster does not need to be changed. Which of the following would be the correct method?

TASK C.4

   A.   Remove an equal amount of shims from both sides.
   B.   Add an equal amount of shims to both sides.
   C.   Add shims to the front and remove an equal amount from the back.
   D.   Remove shims from the front and add an equal amount to the back.

   Answer A is incorrect. This would move the upper ball joint in, moving camber negative.

   Answer B is correct. This would move camber positive without changing caster.

   Answer C is incorrect. This would change caster without changing camber.

   Answer D is incorrect. This would change caster without changing camber.

TASK C.5

39. Which of the following best describes positive caster?

    A. The forward tilt of the steering axis

    B. The tilt of the tire inward

    C. The rearward tilt of the steering axis

    D. The tilt of the tire outward

Answer A is incorrect. This describes negative caster.

Answer B is incorrect. This describes negative camber.

Answer C is correct. This describes positive caster.

Answer D is incorrect. This is the definition of positive camber.

TASK C.4

40. The tires on the front of a vehicle are both worn on the inside edges. Technician A says the vehicle may have too much positive toe. Technician B says the vehicle may have too much positive camber. Who is correct?

    A. A only

    B. B only

    C. Both A and B

    D. Neither A nor B

Answer A is incorrect. Neither Technician is correct. Positive toe would wear the outside edges.

Answer B is incorrect. Neither Technician is correct. Positive camber would wear the outside edges.

Answer C is incorrect. Neither Technician is correct.

Answer D is correct. Neither Technician is correct.

PREPARATION EXAM 5—ANSWER KEY

1.	A	21.	B
2.	D	22.	B
3.	A	23.	A
4.	D	24.	D
5.	A	25.	C
6.	D	26.	C
7.	D	27.	B
8.	C	28.	A
9.	B	29.	D
10.	A	30.	B
11.	A	31.	D
12.	A	32.	B
13.	A	33.	D
14.	A	34.	B
15.	B	35.	D
16.	D	36.	A
17.	C	37.	D
18.	D	38.	D
19.	D	39.	D
20.	C	40.	D

PREPARATION EXAM 5—EXPLANATIONS

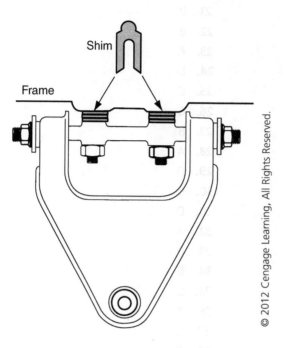

Shim

Frame

© 2012 Cengage Learning, All Rights Reserved.

TASK C.4

1. Camber needs to be moved toward the negative on the suspension system shown in the figure above. Caster does not need to be changed. Which of the following would be the correct method?

   A. Remove an equal amount of shims from both sides.
   B. Add an equal amount of shims to both sides.
   C. Add shims to the front and remove an equal amount from the back.
   D. Remove shims from the front and add an equal amount to the back.

   Answer A is correct. This would move camber negative.

   Answer B is incorrect. This would move camber positive without changing caster.

   Answer C is incorrect. This would change caster without changing camber.

   Answer D is incorrect. This would change caster without changing camber.

TASK C.5

2. Which of the following best describes positive camber?

   A. The forward tilt of the steering axis
   B. The tilt of the tire inward
   C. The rearward tilt of the steering axis
   D. The tilt of the tire outward

   Answer A is incorrect. This describes negative caster.

   Answer B is incorrect. This describes negative camber.

   Answer C is incorrect. This describes positive caster.

   Answer D is correct. This is the definition of positive camber.

3. Which of the following best describes positive toe?

TASK C.7

A. The front of the tires are closer together than the back.

B. The rear of the tires are closer together than the front.

C. The left side of the vehicle has a shorter wheelbase than the right.

D. The right side of the vehicle has a longer wheelbase than the left.

Answer A is correct. This describes positive toe.

Answer B is incorrect. This describes negative toe.

Answer C is incorrect. This describes negative setback.

Answer D is incorrect. This is the definition of positive setback.

4. Which of the following best describes positive setback?

TASK C.14

A. The front of the tires are closer together than the back.

B. The rear of the tires are closer together than the front.

C. The left side of the vehicle has a shorter wheelbase than the right.

D. The right side of the vehicle has a shorter wheelbase than the left.

Answer A is incorrect. This describes positive toe.

Answer B is incorrect. This describes negative toe.

Answer C is incorrect. This describes negative setback.

Answer D is correct. This is the definition of positive setback.

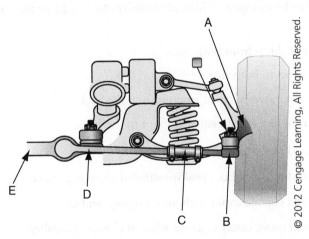

5. Steering axis inclination (SAI) is out of specification on the vehicle shown in the figure above. Which of the following could be the cause?

TASK C.10

A. Item A is bent.

B. Item B is bent.

C. Item C is bent.

D. Item D is bent.

Answer A is correct. A bent steering knuckle (Item A) is a probable cause of incorrect SAI.

Answer B is incorrect. A bent tie rod (Item B) would cause the toe to be incorrect.

Answer C is incorrect. A bent adjusting sleeve (Item C) would cause toe to be incorrect.

Answer D is incorrect. A bent inner tie rod (Item D) would cause toe to be incorrect.

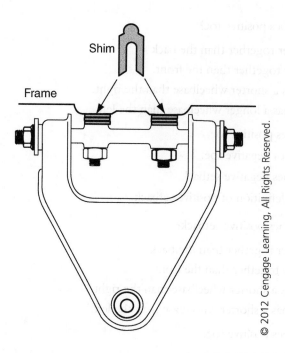

TASK C.6

6. Caster needs to be moved toward the negative on the suspension system shown in the figure above. Camber does not need to be changed. Which of the following would be the correct method?

   A. Remove an equal amount of shims from both sides.

   B. Add an equal amount of shims to both sides.

   C. Add shims to the front and remove an equal amount from the back.

   D. Remove shims from the front and add an equal amount to the back.

   Answer A is incorrect. This would move camber negative.

   Answer B is incorrect. This would move camber positive without changing caster.

   Answer C is incorrect. This would change caster without changing camber.

   Answer D is correct. This would move caster negative without changing camber.

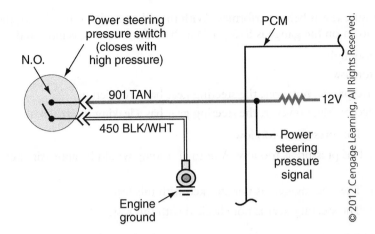

7. A vehicle equipped with the idle compensation system shown in the figure above idles too high at all times. Which of the following could be the cause?

TASK A.2.7

    A. Stuck open pressure switch

    B. Circuit 450 grounded

    C. An open at engine ground

    D. Circuit 901 grounded

Answer A is incorrect. If the switch was stuck open, the idle would not increase.

Answer B is incorrect. Circuit 450 should be grounded.

Answer C is incorrect. An open at the engine ground would prevent the system from idling up the vehicle.

Answer D is correct. If circuit 901 was grounded, this would be a signal to the power train module control (PCM) to idle up the engine.

8. A customer has an occasional hard steering concern which seems to be more common when it is raining. Which of the following is the most likely cause?

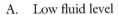

TASK A.2.4

    A. Low fluid level

    B. Faulty power steering pressure switch

    C. Loose power steering pump belt

    D. Low tire pressure

Answer A is incorrect. Low fluid level would not cause steering to be sensitive in rain.

Answer B is incorrect. A faulty power steering pressure switch usually causes engine idle concerns.

Answer C is correct. A loose power steering pump belt will slip more when wet, causing hard steering.

Answer D is incorrect. Low tire pressure would not cause steering to be sensitive in rain.

TASK A.2.7

9. A power steering pressure test is being performed. With the valve on the tester closed, the maximum pressure shown on the gauge is 875 psi. Which of the following is indicated?

A. The pressure is too high.

B. The pressure is too low.

C. The pressure is acceptable, however, the steering gear has a restriction.

D. The pressure is acceptable, however, the steering gear has a leaking internal seal.

Answer A is incorrect. This pressure is too low.

Answer B is correct. This pressure is too low. A normal reading would be approximately 1,500 psi.

Answer C is incorrect. The steering gear is not checked with this test.

Answer D is incorrect. The steering gear is not checked with this test.

TASK A.2.7

10. When a power steering pump pressure test is being performed, what is the maximum amount of time the valve on the tester should be closed?

A. 5 seconds

B. 15 seconds

C. 25 seconds

D. 35 seconds

Answer A is correct. Holding the valve closed longer than 5 seconds can overheat the fluid and damage the pump.

Answer B is incorrect. Holding the valve closed longer than 5 seconds can overheat the fluid and damage the pump.

Answer C is incorrect. Holding the valve closed longer than 5 seconds can overheat the fluid and damage the pump.

Answer D is incorrect. Holding the valve closed longer than 5 seconds can overheat the fluid and damage the pump.

TASK C.5

11. Consider the following front alignment readings:

	Spec	Actual	
		Left	Right
Caster	0.0 degrees	0.0 degrees	0.0 degrees
Camber	1.5 degrees	1.5 degrees	0.5 degrees
Toe	0.5 degrees	0.2 degrees	0.2 degrees
SAI	2.3 degrees	2.2 degrees	2.4 degrees

Which of the following is a true statement?

A. The technician should adjust the camber, then the toe.

B. The technician should check for a bent steering knuckle.

C. The technician should adjust the toe, then camber.

D. The technician should adjust the caster, then camber.

Answer A is correct. The camber adjustment should be performed first, then toe.

Answer B is incorrect. There is no reason to suspect a bent spindle. SAI is acceptable.

Answer C is incorrect. The camber should be adjusted before toe because a camber adjustment will affect the toe, but a toe adjustment will not affect camber.

Answer D is incorrect. The caster does not need to be adjusted on this vehicle.

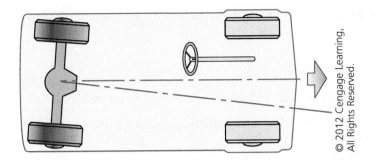

12. The condition shown in the figure above is:

    A.   Positive thrust.

    B.   Negative thrust.

    C.   Positive camber.

    D.   Negative camber.

TASK C.14

    Answer A is correct. This illustrates a positive thrust.

    Answer B is incorrect. This illustrates a positive thrust, not negative.

    Answer C is incorrect. Camber is not illustrated.

    Answer D is incorrect. Camber is not illustrated.

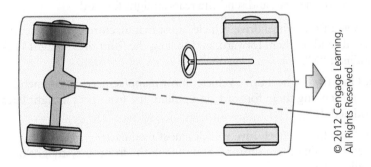

13. The condition shown in the figure above is corrected by:

    A.   Adjusting rear toe.

    B.   Adjusting front toe.

    C.   Adjusting rear camber.

    D.   Adjusting front camber.

TASK C.13

    Answer A is correct. Adjusting rear toe will correct a thrust line problem.

    Answer B is incorrect. Adjusting front toe will not correct a thrust line problem.

    Answer C is incorrect. Adjusting rear camber does not correct a thrust line condition.

    Answer D is incorrect. Adjusting front camber will not correct a thrust line condition.

TASK D.5

14. Which of the following is the normal tire rotational pattern for all-wheel drive vehicles?
    A. Full "X"
    B. Modified "X"
    C. Front to rear
    D. Cross the front, bring to the rear, and send the rears straight forward

Answer A is correct. Most manufacturers recommend crossing the tires in a full "X" when rotating the tires on an all-wheel drive vehicle.

Answer B is incorrect. Most manufacturers recommend crossing the tires in a full "X" when rotating the tires on an all-wheel drive vehicle. A modified "X" involves crossing only the non-drive wheels.

Answer C is incorrect. Most manufacturers recommend crossing the tires in a full "X" when rotating the tires on an all-wheel drive vehicle.

Answer D is incorrect. Most manufacturers recommend crossing the tires in a full "X" when rotating the tires on an all-wheel drive vehicle. This is a modified "X" procedure used on a rear-wheel drive vehicle.

TASK D.5

15. Which of the following is the normal tire rotational pattern for a front-wheel drive vehicle?
    A. Full "X"
    B. Modified "X"
    C. Front to rear
    D. Cross the front, bring to the rear, and send the rears straight forward

Answer A is incorrect. On a front-wheel drive vehicle, most manufacturers recommend crossing the rear tires and sending them forward, and pulling the front tires straight back. This is known as a modified "X".

Answer B is correct. On a front-wheel drive vehicle, most manufacturers recommend crossing the rear tires and sending them forward, and pulling the front tires straight back. This is known as a modified "X".

Answer C is incorrect. On a front-wheel drive vehicle, most manufacturers recommend crossing the rear tires and sending them forward, and pulling the front tires straight back. This is known as a modified "X".

Answer D is incorrect. On a front-wheel drive vehicle, most manufacturers recommend crossing the rear tires and sending them forward, and pulling the front tires straight back. This is known as a modified "X".

16. When checking the tire pressure on a vehicle, where would a technician find the proper air pressure specification?

    A. In the owner's manual
    B. On an under-hood sticker
    C. On an in-trunk sticker
    D. On the driver's door jamb

TASK D.3

    Answer A is incorrect. The specification is usually found on the driver's door jamb.

    Answer B is incorrect. The specification is not on the under-hood sticker.

    Answer C is incorrect. The specification is not found on the in-trunk sticker.

    Answer D is correct. The specification is usually found on the sticker on the driver's door jamb.

17. The tire pressure specification printed on the sidewall of the tire is the:

    A. Recommended cold inflation pressure.
    B. Recommended hot inflation pressure.
    C. Maximum inflation pressure.
    D. Minimum inflation pressure.

TASK D.3

    Answer A is incorrect. The tire pressure specification on the side of the tire is the maximum inflation pressure allowed by the tire manufacturer.

    Answer B is incorrect. Tire pressure is always specified cold.

    Answer C is correct. The tire pressure specification on the side of the tire is the maximum inflation pressure allowed by the tire manufacturer.

    Answer D is incorrect. The tire pressure specification on the side of the tire is the maximum inflation pressure allowed by the tire manufacturer.

18. When should tire pressure be checked?

    A. On a hot tire
    B. After the tire has been driven a minimum of three miles
    C. After the tire has been driven a minimum of three minutes
    D. On a cold tire

TASK D.3

    Answer A is incorrect. Tire pressure should be checked on a cold tire.

    Answer B is incorrect. Tire pressure should be checked on a cold tire.

    Answer C is incorrect. Tire pressure should be checked on a cold tire.

    Answer D is correct. Tire pressure should be checked on a cold tire.

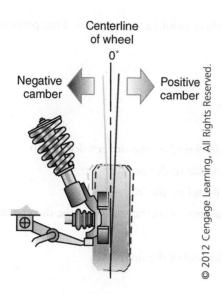

Centerline
of wheel
0°

Negative
camber

Positive
camber

TASK B.1.8

19. Camber needs to be adjusted on the suspension shown in the figure above. Technician A says this can be adjusted by installing new sway bar bushings. Technician B says this can be adjusted by installing new coil springs. Who is correct?

   A. A only

   B. B only

   C. Both A and B

   D. Neither A nor B

   Answer A is incorrect. Neither Technician is correct. Sway bar bushings do not adjust camber.

   Answer B is incorrect. Neither Technician is correct. Although new coil springs will affect camber, they are not installed to adjust camber.

   Answer C is incorrect. Neither Technician is correct.

   Answer D is correct. Neither Technician is correct.

TASK A.2.4

20. A vehicle is hard to steer. Any of these could be the cause EXCEPT:

   A. Seized king pins.

   B. Seized ball joints.

   C. Loose wheel bearings.

   D. Excessive positive caster.

   Answer A is incorrect. Seized king pins can cause tight steering.

   Answer B is incorrect. Seized ball joints can cause tight steering.

   Answer C is correct. Loose wheel bearings would not cause hard steering; they can cause a vehicle to wander.

   Answer D is incorrect. Excessive positive caster can cause a vehicle to be hard to steer.

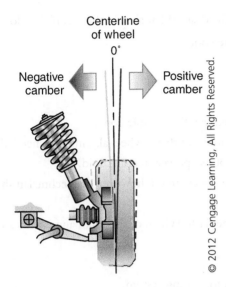

Centerline
of wheel
0°

Negative
camber

Positive
camber

21.　The lower ball joint is being replaced on the suspension system shown in the figure above. Which of the following is correct?

TASK B.1.5

A.　The jack stand must be placed under the lower control arm.

B.　The drive axle must be removed.

C.　The strut must be removed from the vehicle.

D.　The upper strut mount must be removed.

Answer A is incorrect. The jack stand must be placed under the frame.

Answer B is correct. The drive axle must be removed.

Answer C is incorrect. The strut can stay attached to the vehicle.

Answer D is incorrect. The upper strut mount does not need to be removed.

22.　A vehicle pulls to the right. Technician A says the power steering pump could be the cause. Technician B says the power steering gear could be the cause. Who is correct?

TASK C.1

A.　A only

B.　B only

C.　Both A and B

D.　Neither A nor B

Answer A is incorrect. Technician B is correct. Power steering pumps cannot cause a pull. They only provide fluid flow to the steering gear.

Answer B is correct. Only Technician B is correct. Leaking seals within the power steering gear can cause a pull.

Answer C is incorrect. Only Technician B is correct.

Answer D is incorrect. Technician B is correct.

TASK A.1.1

23. The EPS light is continuously illuminated on the dash. What should the technician do?

A. Connect a scan tool and access the trouble code.

B. Align the front axle.

C. Check the power steering fluid level.

D. Center the steering wheel.

Answer A is correct. The technician should retrieve the trouble code.

Answer B is incorrect. The EPS light does not indicate that a wheel alignment is needed.

Answer C is incorrect. The EPS system does not use power steering fluid.

Answer D is incorrect. The steering may need to be centered, however, the technician should retrieve the trouble code first.

TASK A.1.1

24. An icon of a steering wheel is illuminated on the dash. Which of the following is indicated?

A. The power steering fluid is low.

B. The power steering pressure is low.

C. There is a problem with the tire pressure monitoring system.

D. There is a problem in the EPS system.

Answer A is incorrect. The icon does not indicate power steering fluid level.

Answer B is incorrect. The icon does not indicate power steering pressure.

Answer C is incorrect. The icon does not indicate a tire pressure problem.

Answer D is correct. The icon indicates that the EPS system has indentified a malfunction.

TASK A.1.3

25. Technician A says that when reinstalling the clock spring, the steering needs to be in a straight ahead position. Technician B says that when installing a clock spring, the clock spring needs to be centered. Who is correct?

A. A only

B. B only

C. Both A and B

D. Neither A nor B

Answer A is incorrect. Technician B is also correct.

Answer B is incorrect. Technician A is also correct.

Answer C is correct. Both Technicians are correct. The clock spring should be centered and the steering straight ahead to prevent damage to the clock spring when the steering is turned through its normal rotation.

Answer D is incorrect. Both Technicians are correct.

TASK A.3.1

26. The center link on a steering system is unlevel. This would most likely cause:

A. A pull to the left.

B. A pull to the right.

C. Bump steer.

D. Incorrect thrust angle.

Answer A is incorrect. An unlevel center link will not cause a pull.

Answer B is incorrect. An unlevel center link will not cause a pull.

Answer C is correct. An unlevel center link will cause bump steer because as the suspension travels, toe will change, thus effectively self-steering the vehicle.

Answer D is incorrect. An unlevel center link will not change the thrust angle. Rear toe changes the thrust angle.

27. Technician A says a pitman arm should be removed using heat. Technician B says a pitman arm should be removed using a puller. Who is correct?

    A. A only
    B. B only
    C. Both A and B
    D. Neither A nor B

TASK A.3.2

Answer A is incorrect. Technician B is correct. Heat should not be used to remove the pitman arm. This can damage the steering gear and seal as well as change the metallurgy.

Answer B is correct. Only Technician B is correct. A special puller is used to remove the pitman arm.

Answer C is incorrect. Only Technician B is correct.

Answer D is incorrect. Technician B is correct.

28. A vehicle steering wheel is crooked with the vehicle sitting still and the front wheels straight ahead. Which of the following is the most likely cause?

    A. A bent pitman arm
    B. Incorrect thrust line setting
    C. Incorrect thrust angle setting
    D. A bent rear axle

TASK A.3.2

Answer A is correct. A bent pitman arm can cause the steering wheel to be off-center when the front wheels are in a straight ahead position.

Answer B is incorrect. An incorrect thrust line setting could cause the steering wheel to be crooked, but only when the vehicle is rolling.

Answer C is incorrect. An incorrect thrust angle setting could cause the steering wheel to be crooked, but only when the vehicle is rolling.

Answer D is incorrect. A bent rear axle can cause an incorrect thrust angle which could cause the steering wheel to be crooked, but only while the vehicle is rolling.

29. Which of the following could be changed if a vehicle had a bent rear axle?

    A. Front camber
    B. Front toe
    C. Front camber
    D. Thrust angle

TASK B.2.7

Answer A is incorrect. The rear alignment angles would be changed.

Answer B is incorrect. The rear alignment angles would be changed.

Answer C is incorrect. The rear alignment angles would be changed.

Answer D is correct. Thrust angle could be changed because a bent rear axle could change rear toe.

TASK C.12

30. A vehicle with four-wheel steering is being aligned. Which of the following is true?

   A. The steering wheel must be locked before the rear toe can be adjusted.
   B. The rear steering must be locked before the rear toe is adjusted.
   C. The front camber must be set before the rear toe.
   D. The front toe must be set before rear toe.

   Answer A is incorrect. The steering wheel does not need to be locked before rear toe is adjusted; it must be locked before front toe is adjusted.

   Answer B is correct. The rear steering must be locked in the center prior to setting the rear toe.

   Answer C is incorrect. The front camber would be adjusted after the rear of the vehicle is adjusted.

   Answer D is incorrect. The front toe is the last adjustment.

TASK B.2.4

31. During a pre-alignment inspection, the technician finds rust streaks on one rear leaf spring. Which of the following should be performed?

   A. The spring should be primed, then painted.
   B. The spring should be replaced.
   C. Both rear springs should be replaced.
   D. The spring should be inspected for cracks.

   Answer A is incorrect. The spring may be cracked and should be inspected.

   Answer B is incorrect. The spring needs to be inspected to determine if replacement is necessary.

   Answer C is incorrect. If the spring on one side is replaced, then the other side will need to be replaced. However, the determination to replace the spring has not been made; the spring needs to be inspected.

   Answer D is correct. Rust streaks can indicate cracks. The spring needs to be inspected.

TASK B.2.4

32. There are rust streaks on the rear leaf spring where it contacts the rear axle. Which of the following is true?

   A. The spring must be replaced.
   B. The spring center bolt may be sheared.
   C. The rear camber needs to be adjusted.
   D. The rear axle must be replaced.

   Answer A is incorrect. The spring needs to be inspected.

   Answer B is correct. The spring center bolt may be sheared; this would allow the axle to move on the spring.

   Answer C is incorrect. The rear camber may be off, however, the camber is not the cause of the movement between the axle and pad.

   Answer D is incorrect. The rear axle must be inspected to determine if replacement is necessary.

33. An upper strut mount needs to be replaced on a double wishbone front suspension. Technician A says the strut must also be replaced. Technician B says the coil spring must also be replaced. Who is correct?

    A. A only

    B. B only

    C. Both A and B

    D. Neither A nor B

    Answer A is incorrect. Neither Technician is correct. The strut does not have to be replaced unless it fails inspection.

    Answer B is incorrect. Neither Technician is correct. The coil spring does not need to be replaced unless it fails inspection.

    Answer C is incorrect. Neither Technician is correct.

    Answer D is correct. Neither Technician is correct.

TASK B.1.13

34. A vehicle with torsion bar front suspension is low on the left front corner. Which of the following is the correct repair procedure?

    A. Attempt to adjust the right front torsion bar.

    B. Attempt to adjust the left front torsion bar.

    C. Replace the left front torsion bar.

    D. Replace both torsion bars.

    Answer A is incorrect. The left front is low, therefore, an attempt should be made to adjust the left front, not the right front.

    Answer B is correct. An attempt to adjust the left front torsion bar should be made to correct ride height.

    Answer C is incorrect. The left front torsion bar should be adjusted, if possible.

    Answer D is incorrect. The left torsion bar should be adjusted.

TASK B.1.10

35. The torsion bars on a vehicle have been removed and are being reinstalled. Which of the following is correct concerning reinstallation?

    A. The torsion bars must be replaced.

    B. The torsion bars should be reinstalled on the opposite side from removal to equalize wear.

    C. The torsion bars should be turned end-for-end to equalize wear.

    D. The final adjustment should be in the upward direction.

    Answer A is incorrect. The bars do not need to be replaced.

    Answer B is incorrect. The torsion bars should be reinstalled in their original location.

    Answer C is incorrect. The torsion bars should be reinstalled in their original location.

    Answer D is correct. The final adjustment should be in the upward direction. This helps prevent settling of the vehicle after it is put back into service.

TASK B.1.10

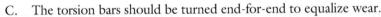

36. A right rear lower ball joint is being replaced on an independent rear suspension. Where should the jack stand be placed during this procedure?

    A. Under the frame on the right side
    B. Under the right rear lower control arm
    C. Under the frame on the left side
    D. Under the left rear control arm

    Answer A is correct. The right side of the frame must be supported. The suspension needs to be unloaded.

    Answer B is incorrect. The right side of the frame must be supported. Installing the jack stand under the right rear lower control arm would keep the ball joint loaded.

    Answer C is incorrect. The right side of the frame of the frame must be supported.

    Answer D is incorrect. The right side of the frame must be supported. The right lower ball joint is being replaced, not the left.

Shim

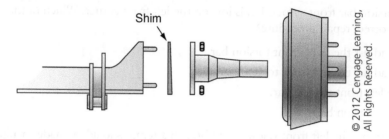

37. A vehicle equipped with the suspension system shown in the figure above is having the spindle replaced. The shim being installed as shown above will:

    A. Correct axle length.
    B. Correct spindle length.
    C. Correct toe.
    D. Correct camber.

    Answer A is incorrect. The shim being installed as shown would correct camber.

    Answer B is incorrect. The shim being installed as shown would correct camber.

    Answer C is incorrect. The shim being installed as shown would correct camber because the taper is shown top to bottom.

    Answer D is correct. The shim being installed as shown would correct camber; in this position it would cause the camber to be moved negative.

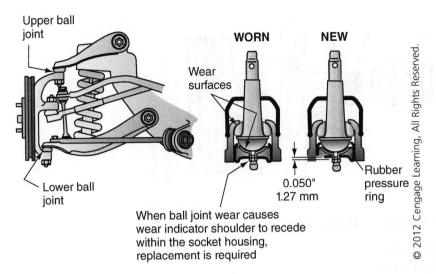

Upper ball joint

Lower ball joint

WORN   NEW

Wear surfaces

When ball joint wear causes wear indicator shoulder to recede within the socket housing, replacement is required

0.050"
1.27 mm

Rubber pressure ring

38. To inspect the lower ball joint on the suspension system shown in the figure above, the technician should:

TASK B.1.5

A. Place a jack stand under the frame.
B. Place a jack stand under the lower control arm.
C. Support the vehicle on a frame hoist.
D. Inspect the joints with the vehicle setting on the tires.

Answer A is incorrect. This is a wear indicator ball joint. The joint should be inspected with the weight of the vehicle on the tires.

Answer B is incorrect. This is a wear indicator ball joint. The joint should be inspected with the weight of the vehicle on the tires.

Answer C is incorrect. This is a wear indicator ball joint. The joint should be inspected with the weight of the vehicle on the tires.

Answer D is correct. This is a wear indicator ball joint. The joint should be inspected with the weight of the vehicle on the tires.

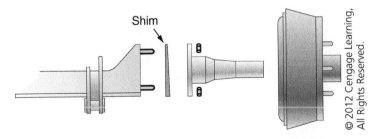

Shim

39. The shim is being installed as shown in the figure above. This will cause which of the following?

TASK C.4

A. Toe to move positive
B. Toe to move negative
C. Camber to move positive
D. Camber to move negative

Answer A is incorrect. To move toe positive, the taper should be front to rear.

Answer B is incorrect. To move toe negative, the taper should be rear to front.

Answer C is incorrect. If the shim is installed upside down, it will cause camber to go positive.

Answer D is correct. The installation as shown will cause camber to go negative.

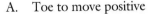

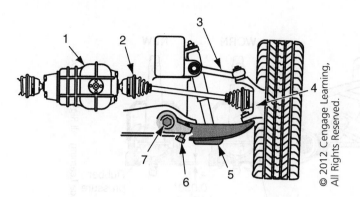

TASK C.2

40. Item #6 in the figure above is used to:

A. Adjust camber.

B. Adjust caster.

C. Adjust toe.

D. Adjust ride height.

Answer A is incorrect. This is the torsion bar adjustment to adjust ride height. Camber is usually adjusted at the upper control arm.

Answer B is incorrect. This is the torsion bar adjustment to adjust ride height. Caster is usually adjusted at the upper control arm.

Answer C is incorrect. This is the torsion bar adjustment to adjust ride height. Toe is adjusted with the tie rod end.

Answer D is correct. This is the torsion bar adjustment to adjust ride height.

PREPARATION EXAM 6—ANSWER KEY

1.	C	21.	B
2.	A	22.	D
3.	A	23.	C
4.	D	24.	D
5.	B	25.	B
6.	A	26.	C
7.	D	27.	B
8.	A	28.	B
9.	D	29.	B
10.	C	30.	C
11.	A	31.	A
12.	D	32.	B
13.	D	33.	C
14.	B	34.	B
15.	C	35.	C
16.	D	36.	C
17.	B	37.	A
18.	A	38.	D
19.	B	39.	A
20.	C	40.	B

PREPARATION EXAM 6—EXPLANATIONS

1. When should the power steering fluid be checked?

    A. Only when the fluid is hot
    B. Only when the fluid is cold
    C. When the fluid is hot or cold
    D. When the steering is straight ahead

TASK A.2.3

Answer A is incorrect. The dipstick is marked with both hot and cold level marks.

Answer B is incorrect. The dipstick is marked with both hot and cold level marks.

Answer C is correct. The dipstick is marked with both hot and cold level marks.

Answer D is incorrect. The position of the steering is not important.

2. A power steering pump makes noise all the time. Which of the following is the most likely cause?

    A. Aerated fluid
    B. Stuck open pressure relief valve
    C. Stuck closed pressure relief valve
    D. Slightly loose belt

Answer A is correct. Aerated fluid can cause a power steering pump to be noisy all the time.

Answer B is incorrect. A stuck open pressure relief valve would cause low power steering assist.

Answer C is incorrect. A stuck closed pressure relief valve would cause high pressure when the steering wheel was turned completely left or right.

Answer D is incorrect. A slightly loose belt may cause noise, however, it would most likely be when the power steering pressure demand was high, such as during a parking maneuver.

3. Which of the following is the most likely cause of a tire which is worn on both outside shoulders?

    A. Underinflation
    B. Overinflation
    C. Positive caster
    D. Positive camber

Answer A is correct. Underinflation causes wear on both of the outside shoulders.

Answer B is incorrect. Overinflation causes wear in the center.

Answer C is incorrect. Caster is not a tire wear angle.

Answer D is incorrect. Positive camber would cause tire wear on the outer shoulder.

4. A tire has a leak in the sidewall. Which of the following is the correct repair procedure?

    A. Locate the source of the leak and plug it from the outside.
    B. Locate the source of the leak and plug it from the inside.
    C. Locate the source of the leak, plug it from the outside, and patch it from the inside.
    D. Replace the tire.

Answer A is incorrect. When a tire is losing air through the sidewall, the recommended procedure is to replace the tire.

Answer B is incorrect. When a tire is losing air through the sidewall, the recommended procedure is to replace the tire.

Answer C is incorrect. When a tire is losing air through the sidewall, the recommended procedure is to replace the tire.

Answer D is correct. When a tire is losing air through the sidewall, the recommended procedure is to replace the tire.

5. A customer is concerned that it takes increasing effort to steer the car. Which of the following could be the cause?

   A. Worn steering column bushings
   B. Seized steering column U-joint
   C. Worn tie rod end
   D. Loose idler arm

TASK A.1.1

Answer A is incorrect. When steering column bushings wear, they will normally create less friction not more. The usual customer complaint is a noisy steering column.

Answer B is correct. When the bearings in a U-joint fail, they will often "freeze." This will result in a binding steering system, as the needle bearings will not travel around the trunnion normally. This problem is often misdiagnosed as a weak power steering pump.

Answer C is incorrect. When a tie rod wears, they usually develop excess movement. This results in looseness in the steering system, noise, and tie wear.

Answer D is incorrect. A worn idler arm will produce looseness in the steering system causing the vehicle to wander. The customer may comment that they need to continually correct the steering to keep the vehicle on the road.

6. A car equipped with power steering pulls (leads) to the left after a proper alignment. Which of the following is the most likely cause?

   A. A power steering gear
   B. Power steering pump
   C. Pitman arm
   D. All four tires overinflated

TASK A.2.1

Answer A is correct. A power steering gear can cause a vehicle to pull or lead. To diagnose this concern, jack the vehicle up and secure on jack stands with the tires in the straight ahead position. Start the vehicle and watch for the front wheels to self-steer. If the steering system turns the wheels on its own, the steering gear box is faulty.

Answer B is incorrect. A faulty power steering pump will cause increased steering effort but will not cause a pull or lead.

Answer C is incorrect. A worn pitman arm will cause the steering system to have excess play which will create a loose feeling in the steering.

Answer D is incorrect. If all the tires were overinflated, this would cause tread wear in the center of the tires, not a pull.

TASK A.2.15

7. During a dry park inspection, the technician finds power steering fluid leaking from the rack and pinion bellows. Which of the following is the most likely cause?

    A. An incorrectly installed bellows
    B. A missing bellows clamp
    C. Excessive power steering pump pressure
    D. Worn rack seals

    Answer A is incorrect. There should not be oil in the rack bellows. The bellows are used to prevent dirt contamination of the inner tie rod end. If the bellows were installed incorrectly, the most likely result would be a worn inner tie rod end due to dirt contamination.

    Answer B is incorrect. A missing bellows clamp would allow dirt to enter the inner tie rod end, resulting in inner tie rod end wear.

    Answer C is incorrect. Excessive power steering pump pressure would result in sensitive steering and premature power steering pump failure.

    Answer D is correct. Worn rack seals would allow oil to leak past the seals and enter the bellows. When this condition occurs, the normal repair procedure is to replace the rack.

TASK A.2.1

8. Which of the following is the correct inspection procedure for an idler arm?

    A. With the wheels on the ground, grasp the idler arm and attempt to move it with hand pressure only.
    B. With the wheels on the ground, use a large pair of pliers to attempt to move the idler arm.
    C. Lift the vehicle on a frame hoist, grasp the idler arm, and attempt to move it with hand pressure only.
    D. Lift the vehicle on a frame hoist and turn the steering from side to side looking for play.

    Answer A is correct. The wheels should be on the ground and hand pressure should be used to check the idler arm for excess movement.

    Answer B is incorrect. Pliers should not be used, as they would apply too much force on the joint and cause it to fail the inspection, resulting in an unnecessary parts replacement.

    Answer C is incorrect. The inspection of steering system components is always performed at dry park.

    Answer D is incorrect. The inspection of steering system components is always performed at dry park.

TASK A.2.1

9. A vehicle has steering wheel shimmy after crossing a railroad track. Which of the following is the most likely cause?

    A. Worn tie rod end
    B. Worn ball joint
    C. Weak front springs
    D. Worn steering damper

    Answer A is incorrect. A worn tie rod end will result in steering system looseness and tire wear.

    Answer B is incorrect. Worn ball joints will result in looseness and tire wear.

    Answer C is incorrect. Weak front springs will result in lower than normal ride height.

    Answer D is correct. A worn steering damper will allow the steering system to turn past center when coming out of a turn and result in a shimmy condition.

10. A vehicle has worn front and rear jounce bumpers. All of these could be the cause EXCEPT:

    A. Worn shocks.

    B. Weak springs.

    C. Worn wheel bearings.

    D. Incorrect ride height.

TASK B.1.3, B.2.5

    Answer A is incorrect. Worn shocks will cause the suspension to travel further than normal, resulting in worn jounce bumpers.

    Answer B is incorrect. Weak springs will allow the vehicle to set lower than normal and result in worn jounce bumpers.

    Answer C is correct. Worn wheel bearings can result in noise, but would not cause suspension travel to be abnormal.

    Answer D is incorrect. Incorrect ride height could cause the vehicle to set lower than normal, resulting in worn jounce bumpers.

11. Both front springs have been replaced on the vehicle. Which of the following must also be performed?

    A. Align the vehicle

    B. Replace the rear springs

    C. Replace the shocks

    D. Perform a power steering pressure test

TASK B.1.8

    Answer A is correct. After the springs are replaced, drive the vehicle to settle the suspension into its normal ride height and then perform an alignment.

    Answer B is incorrect. The rear springs should only be replaced if they are found to be faulty.

    Answer C is incorrect. The shocks do not need to be replaced unless they are worn.

    Answer D is incorrect. There is no reason to perform a power steering pressure test just because the front springs have been replaced.

12. A strut bearing is noisy and must be replaced. Technician A says to replace the strut cartridge while replacing the strut bearing. Technician B says to replace the front coil spring while replacing the strut bearing. Who is correct?

    A. A only

    B. B only

    C. Both A and B

    D. Neither A nor B

TASK B.1.13

    Answer A is incorrect. Neither Technician is correct. The strut cartridge should only be replaced if it is worn.

    Answer B is incorrect. Neither Technician is correct. The front coil spring should only be replaced if it is worn.

    Answer C is incorrect. Neither Technician is correct.

    Answer D is correct. Neither Technician is correct.

TASK B.2.2

13. A vehicle sits lower on the left rear than on the right rear. Which of the following is the most likely cause?

A. The left rear shock is weak.

B. The right shock was replaced and the left shock was not.

C. The lateral link bushings are worn.

D. The left rear spring is weak.

Answer A is incorrect. Shocks do not control ride height.

Answer B is incorrect. Shocks do not control ride height.

Answer C is incorrect. Neither the lateral link nor its bushings control ride height. They control side to side movement of the rear axle.

Answer D is correct. A weak left rear spring could cause the vehicle to set low on the left rear.

TASK B.2.3

14. A vehicle has worn sway bar end link bushings. Which of the following would be the most likely customer complaint?

A. Noise while driving in a straight ahead position on a smooth road

B. Noise while driving on an uneven road

C. Lower than normal ride height

D. Tire wear on the outside of the tread

Answer A is incorrect. A sway bar does not move while driving in a straight ahead position on a smooth road.

Answer B is correct. The sway bar moves while driving on an uneven road or during a turn. This is when the worn sway bar end link bushings would generate a noise.

Answer C is incorrect. Sway bars and sway bar end link bushings do not affect ride height.

Answer D is incorrect. Sway bars and sway bar end link bushings do not cause tire wear, they cause vehicle handling concerns.

TASK B.2.4

15. A composite leaf spring is being replaced on a vehicle. Which of the following is a true statement about this procedure?

A. The spring should be lubricated prior to installation.

B. The spring must first be compressed in a shop press.

C. Care should be taken not to scratch the spring.

D. The spring can be installed with either side up.

Answer A is incorrect. The spring does not need to be lubricated prior to installation.

Answer B is incorrect. The spring must be installed with a spring compressor designed for that purpose, not a shop press.

Answer C is correct. The spring should not be scratched or damaged in any way. This will weaken the spring and cause premature failure.

Answer D is incorrect. The spring can only be installed with the proper side up.

16. A vehicle equipped with air ride rear suspension will not lower when a load is removed from the trunk. Which of the following could be the cause?

TASK C.3

   A. A plugged air supply solenoid

   B. A worn air compressor

   C. A faulty air compressor relay

   D. A plugged air vent solenoid

   Answer A is incorrect. A plugged air supply solenoid would result in a vehicle that would not raise because air would not enter the suspension.

   Answer B is incorrect. A worn air compressor would not be able to produce enough pressure to raise the vehicle.

   Answer C is incorrect. A faulty air compressor relay would cause the compressor to fail to run and that would cause the vehicle to set too low.

   Answer D is correct. A plugged air vent solenoid would prevent air from escaping from the air suspension, resulting in a vehicle that would not return to normal ride height after a load was removed from the trunk.

17. Vehicle ride height is correct on the front, but lower than specification on the rear. Which of the following front alignment angles would be most affected?

TASK C.2

   A. Toe

   B. Caster

   C. Camber

   D. SAI

   Answer A is incorrect. Front toe angle would not be greatly affected by rear ride height. However, front ride height would affect front toe.

   Answer B is correct. Caster is the forward and rearward tilt of the steering axis. When rear ride height is lower than specification, it has the effect of making caster more positive.

   Answer C is incorrect. Camber is the inward or outward tilt of the tire. Rear ride height has little effect on front camber.

   Answer D is incorrect. SAI is included angle plus camber. Rear ride height would have little effect on this angle.

18. Which alignment angle is adjusted last during a four-wheel alignment?

TASK C.7

   A. Front toe

   B. Rear toe

   C. Front camber

   D. Rear camber

   Answer A is correct. The correct sequence of adjust during a four-wheel alignment is rear camber, rear toe, front camber/caster, then front toe.

   Answer B is incorrect. The correct sequence of adjust during a four-wheel alignment is rear camber, rear toe, front camber/caster, then front toe.

   Answer C is incorrect. The correct sequence of adjust during a four-wheel alignment is rear camber, rear toe, front camber/caster, then front toe.

   Answer D is incorrect. The correct sequence of adjust during a four-wheel alignment is rear camber, rear toe, front camber/caster, then front toe.

TASK C.9

19. Toe-out-on-turns is not within specification. Which of the following is the most likely cause?

    A. Bent spindle

    B. Bent steering arm

    C. Bent strut

    D. Bent shock

Answer A is incorrect. A bent spindle would change SAI, caster, or camber but will not affect toe-out-on-turns.

Answer B is correct. A bent steering arm will cause toe-out-on-turns to be incorrect.

Answer C is incorrect. A bent strut will change SAI, caster, or camber but will not affect toe-out- on-turns.

Answer D is incorrect. A bent shock will not change alignment angles.

TASK C.12

20. Technician A says rear toe on some vehicles can be adjusted with shims. Technician B says rear toe on some vehicles can be adjusted with cams. Who is correct?

    A. A only

    B. B only

    C. Both A and B

    D. Neither A nor B

Answer A is incorrect. Technician B is also correct.

Answer B is incorrect. Technician A is also correct.

Answer C is correct. Both Technicians are correct. Both shims and cams can be used to adjust rear toe.

Answer D is incorrect. Both Technicians are correct.

TASK C.13

21. A four-wheel alignment is being performed on a vehicle. Thrust angle needs to be corrected. Which adjustment is made to adjust the thrust angle?

    A. Rear camber

    B. Rear toe

    C. Front camber

    D. Front toe

Answer A is incorrect. Rear camber is not the correct way to adjust thrust angle.

Answer B is correct. Thrust angle is the angle formed by the thrust line and the geometric center line. In order to adjust thrust angle, the technician should adjust the rear toe.

Answer C is incorrect. Front camber will not correct the thrust angle.

Answer D is incorrect. Front toe will not correct the thrust angle.

22. A tire has the following marked on the sidewall: P235/75R16. Which of the following is true?

    A. The tire has a section width of 75 mm.

    B. The tire has an aspect ratio of 235.

    C. The tire has been re-grooved.

    D. The tire fits a 16 inch diameter rim.

TASK D.2

Answer A is incorrect. The section width is 235 mm.

Answer B is incorrect. The aspect ratio is 75.

Answer C is incorrect. The R indicates that the tire is a radial.

Answer D is correct. The tire does fit a 16 inch diameter rim.

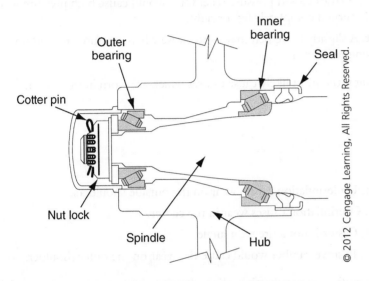

23. The wheel bearings shown in the figure above are being serviced. Which of the following is true concerning the correct tightening procedure?

    A. Tighten the nut to 100 ft lbs.

    B. Tighten the nut to specification and back off two turns.

    C. The final check should be made with a dial indicator.

    D. The wheels/tires must be on the ground.

TASK C.2

Answer A is incorrect. 100 ft lbs. would apply too much preload to the bearing.

Answer B is incorrect. This would allow too much freeplay.

Answer C is correct. The final check should be 0.001"–0.005" (0.025 mm–0.127 mm).

Answer D is incorrect. The wheel end must be rotated during the tightening procedure.

TASK A.1.1

24. A power steering pump makes noise while turning the steering wheel. Which of the following is the most likely cause?

   A. Aerated fluid

   B. Stuck open pressure relief valve

   C. Stuck closed pressure relief valve

   D. Slightly loose belt

Answer A is incorrect. Aerated fluid can cause a power steering pump to be noisy all the time.

Answer B is incorrect. A stuck open pressure relief valve would cause low power steering assist.

Answer C is incorrect. A stuck closed pressure relief valve would cause high pressure when the steering wheel was turned completely left or right.

Answer D is correct. A slightly loose belt may cause noise when pressure demand was high, such as during a parking maneuver.

TASK D.1

25. Which of the following is the most likely cause of tires which are worn in the center?

   A. Underinflation

   B. Overinflation

   C. Positive caster

   D. Positive camber

Answer A is incorrect. Underinflation causes wear on the outside shoulders.

Answer B is correct. Overinflation causes wear in the center.

Answer C is incorrect. Caster is not a tire wear angle.

Answer D is incorrect. Positive camber would cause tire wear on the outer shoulder.

TASK C.9

26. After setting toe to manufacture specification, the technician checks toe-out-on-turns. When the front left wheel is turned out 20 degrees, the right front wheel turns in 24 degrees. What does this indicate?

   A. A normal operating steering system

   B. A bent strut

   C. A bent steering arm

   D. A worn ball joint

Answer A is incorrect. Normal readings when checking toe-out-on-turns is closer to 20 degrees and 18 degrees.

Answer B is incorrect. A bent strut would not change this measurement.

Answer C is correct. A bent steering arm is indicated.

Answer D is incorrect. A worn ball joint would not change toe-out-on-turns.

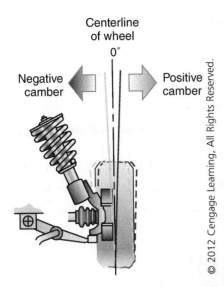

Centerline
of wheel
0°

Negative
camber

Positive
camber

27. The suspension system shown in the figure above is being inspected. Technician A says the load carrying joint is the lower ball joint. Technician B says the load carry joint is the upper strut mount. Who is correct?

TASK B.1.13

   A. A only

   B. B only

   C. Both A and B

   D. Neither A nor B

   Answer A is incorrect. Technician B is correct. The lower ball joint is the follower joint on this suspension system.

   Answer B is correct. Only Technician B is correct. The load is carried in the upper strut mount.

   Answer C is incorrect. Only Technician B is correct.

   Answer D is incorrect. Technician B is correct.

28. A vehicle with conventional (recirculating ball) steering has a diagnostic code for a faulty steering wheel position sensor. Where is the most likely location for the sensor?

TASK A.1.2

   A. At the base of the steering column

   B. Behind the steering wheel

   C. On the steering gear box

   D. Inside the steering gear box

   Answer A is incorrect. The most likely location is behind the steering wheel, not the base of the steering column.

   Answer B is correct. The most likely location is behind the steering wheel.

   Answer C is incorrect. The most likely location is behind the steering wheel, not on the steering gear box.

   Answer D is incorrect. The most likely location is behind the steering wheel, not inside the steering gear box.

TASK C.8

29. When reinstalling the steering wheel position sensor:

    A.   The vehicle alignment must be adjusted.

    B.   The sensor may need to be centered.

    C.   The power steering system will need to be bled.

    D.   The power steering pressure will need to be checked.

Answer A is incorrect. Replacing the steering wheel position sensor will not change the vehicle alignment.

Answer B is correct. Some models will require the sensor to be centered when installed.

Answer C is incorrect. Replacing the sensor will not introduce air into the system.

Answer D is incorrect. The sensor does not affect the power steering pressure.

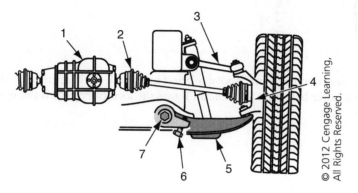

TASK B.1.2

30. The pivot bar bushings in item #3 shown in the figure above need to be replaced. Which of the following is correct?

    A.   Item #2 must be removed.

    B.   Item #5 must be removed.

    C.   The jack stand should be placed under item #5.

    D.   Item #6 should be removed.

Answer A is incorrect. Item #2 is the drive axle; it does not need to be removed.

Answer B is incorrect. Item #5 is the lower control arm; it does not need to be removed.

Answer C is correct. The jack stand should be placed under item #5.

Answer D is incorrect. The torsion bar adjusting bolt, item #6, does not need to be removed.

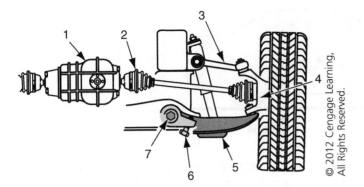

31. Camber needs to be adjusted on the suspension system shown in the figure above. This would most likely occur at:

    A. #3.

    B. #4.

    C. #5.

    D. #6.

TASKS C.4, B.1.2

Answer A is correct. The upper control arm (#3) is the most likely place to adjust camber on this suspension system.

Answer B is incorrect. The steering knuckle (#4) does not usually contain the camber adjustment on this type of suspension.

Answer C is incorrect. The lower control arm (#5) may have a camber adjustment feature, however, it is not the most likely to need adjustment.

Answer D is incorrect. Item #6 is not the camber adjustment; it is the torsion bar adjustment.

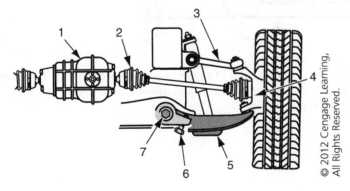

32. Technician A says the suspension system shown in the figure above is a McPherson strut. Technician B says the suspension system shown in the figure above is adjustable. Who is correct?

    A. A only

    B. B only

    C. Both A and B

    D. Neither A nor B

TASK B.1.10

Answer A is incorrect. Technician B is correct. This is a SLA suspension.

Answer B is correct. Only Technician B is correct. This suspension has an adjustable torsion bar.

Answer C is incorrect. Only Technician B is correct.

Answer D is incorrect. Technician B is correct.

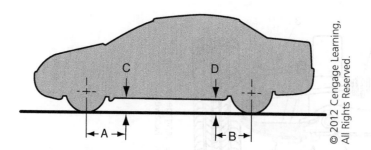

TASK B.2.1

33. In the figure shown above, dimension D is less than specification. Dimension C is within specification. Which of the following is the correct repair procedure?

    A. Replace all four springs.

    B. Replace the front two springs.

    C. Replace the rear two springs.

    D. Replace the springs on the left side of the vehicle.

Answer A is incorrect. Only the rear springs need to be replaced.

Answer B is incorrect. The front is not low.

Answer C is correct. The rear springs should be replaced in a pair.

Answer D is incorrect. Springs are replaced across an axle, not on a side.

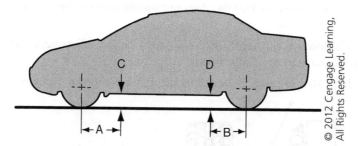

TASK B.1.1

34. In the figure shown above, dimensions C and D are greater than specification. Technician A says all four springs must be replaced. Technician B says this could be caused by using aftermarket tires and wheels. Who is correct?

    A. A only

    B. B only

    C. Both A and B

    D. Neither A nor B

Answer A is incorrect. Technician B is correct. It is very unlikely that the springs need to be replaced.

Answer B is correct. Only Technician B is correct. When a vehicle sets higher than specification, the most likely cause is that the tire/wheel combination is taller than the originals.

Answer C is incorrect. Only Technician B is correct.

Answer D is incorrect. Technician B is correct.

35. Included angle is the combination of:

    A. Camber and caster.
    B. Front and rear toe.
    C. SAI and camber.
    D. SAI and caster.

TASK C.10

Answer A is incorrect. Included angle is the combination of SAI and camber.

Answer B is incorrect. Included angle is the combination of SAI and camber.

Answer C is correct. Included angle is the combination of SAI and camber.

Answer D is incorrect. Included angle is the combination of SAI and camber.

36. A vehicle has a worn tie rod end on the left front. Which of the following would be affected?

    A. Rear toe
    B. Front camber
    C. Front toe
    D. Front caster

TASK C.7

Answer A is incorrect. A front tie rod would not affect rear toe.

Answer B is incorrect. A front tie rod would not affect front camber.

Answer C is correct. A front tie rod would affect front toe.

Answer D is incorrect. A front tie rod would not affect front caster.

37. A vehicle has a worn tie rod end on the left rear. Which of the following would be affected?

    A. Rear toe
    B. Rear camber
    C. Front toe
    D. Front caster

TASK C.12

Answer A is correct. A rear tie rod would affect rear toe.

Answer B is incorrect. A rear tie rod would not affect rear camber.

Answer C is incorrect. A rear tie rod would not affect front toe.

Answer D is incorrect. A front tie rod would not affect front caster.

38. Technician A says camber pulls to the least positive. Technician B says caster pulls to the most positive. Who is correct?

    A. A only
    B. B only
    C. Both A and B
    D. Neither A nor B

TASK C.1

Answer A is incorrect. Camber causes the tire to roll like a cone. Camber pulls to the most positive.

Answer B is incorrect. Caster pulls to the least positive.

Answer C is incorrect. Neither Technician is correct.

Answer D is correct. Neither Technician is correct.

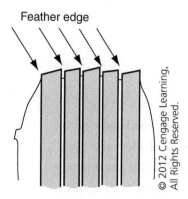

Feather edge

TASK D.1

39. Which of the following is the most likely cause of the tire wear shown in the figure above?

    A. Toe

    B. Camber

    C. Caster

    D. SAI

Answer A is correct. This is a front toe wear pattern.

Answer B is incorrect. This is not a camber wear. Camber wears the inner or outer tread.

Answer C is incorrect. Caster is not a tire wear angle.

Answer D is incorrect. SAI only indirectly affects tire wear, similar to a camber wear pattern. This is a toe wear pattern.

TASK C.1

40. The center link needs to be replaced on a vehicle. Technician A says the outer tie rods will also need to be replaced. Technician B says the toe will need to be adjusted. Who is correct?

    A. A only

    B. B only

    C. Both A and B

    D. Neither A nor B

Answer A is incorrect. Technician B is correct. The outer tie rods do not have to be replaced with the center link.

Answer B is correct. Only Technician B is correct. Toe will need to be adjusted after the center link is replaced.

Answer C is incorrect. Only Technician B is correct.

Answer D is incorrect. Technician B is correct.

Appendices

PREPARATION EXAM ANSWER SHEET FORMS

ANSWER SHEET

1. _____
2. _____
3. _____
4. _____
5. _____
6. _____
7. _____
8. _____
9. _____
10. _____
11. _____
12. _____
13. _____
14. _____
15. _____
16. _____
17. _____
18. _____
19. _____
20. _____

21. _____
22. _____
23. _____
24. _____
25. _____
26. _____
27. _____
28. _____
29. _____
30. _____
31. _____
32. _____
33. _____
34. _____
35. _____
36. _____
37. _____
38. _____
39. _____
40. _____

ANSWER SHEET

1.	_____	21.	_____
2.	_____	22.	_____
3.	_____	23.	_____
4.	_____	24.	_____
5.	_____	25.	_____
6.	_____	26.	_____
7.	_____	27.	_____
8.	_____	28.	_____
9.	_____	29.	_____
10.	_____	30.	_____
11.	_____	31.	_____
12.	_____	32.	_____
13.	_____	33.	_____
14.	_____	34.	_____
15.	_____	35.	_____
16.	_____	36.	_____
17.	_____	37.	_____
18.	_____	38.	_____
19.	_____	39.	_____
20.	_____	40.	_____

ANSWER SHEET

1. _____
2. _____
3. _____
4. _____
5. _____
6. _____
7. _____
8. _____
9. _____
10. _____
11. _____
12. _____
13. _____
14. _____
15. _____
16. _____
17. _____
18. _____
19. _____
20. _____

21. _____
22. _____
23. _____
24. _____
25. _____
26. _____
27. _____
28. _____
29. _____
30. _____
31. _____
32. _____
33. _____
34. _____
35. _____
36. _____
37. _____
38. _____
39. _____
40. _____

ANSWER SHEET

1. _____	21. _____
2. _____	22. _____
3. _____	23. _____
4. _____	24. _____
5. _____	25. _____
6. _____	26. _____
7. _____	27. _____
8. _____	28. _____
9. _____	29. _____
10. _____	30. _____
11. _____	31. _____
12. _____	32. _____
13. _____	33. _____
14. _____	34. _____
15. _____	35. _____
16. _____	36. _____
17. _____	37. _____
18. _____	38. _____
19. _____	39. _____
20. _____	40. _____

Delmar, Cengage Learning ASE Test Preparation

ANSWER SHEET

1. _____

2. _____

3. _____

4. _____

5. _____

6. _____

7. _____

8. _____

9. _____

10. _____

11. _____

12. _____

13. _____

14. _____

15. _____

16. _____

17. _____

18. _____

19. _____

20. _____

21. _____

22. _____

23. _____

24. _____

25. _____

26. _____

27. _____

28. _____

29. _____

30. _____

31. _____

32. _____

33. _____

34. _____

35. _____

36. _____

37. _____

38. _____

39. _____

40. _____

ANSWER SHEET

1.	_____	21.	_____
2.	_____	22.	_____
3.	_____	23.	_____
4.	_____	24.	_____
5.	_____	25.	_____
6.	_____	26.	_____
7.	_____	27.	_____
8.	_____	28.	_____
9.	_____	29.	_____
10.	_____	30.	_____
11.	_____	31.	_____
12.	_____	32.	_____
13.	_____	33.	_____
14.	_____	34.	_____
15.	_____	35.	_____
16.	_____	36.	_____
17.	_____	37.	_____
18.	_____	38.	_____
19.	_____	39.	_____
20.	_____	40.	_____

Glossary

Actuator A device that delivers motion in response to an electrical signal.

A/D Converter Abbreviation for Analog-to-Digital Converter.

Additive An additive intended to improve a certain characteristic of the material or fluid.

Aftercooler A charge air cooling device, usually water-cooled.

Air Compressor An engine-driven mechanism for supplying high pressure air to the truck brake system.

Air Filter A device that minimizes the possibility of impurities entering the intake system.

Altitude Compensation System An altitude barometric switch and solenoid used to provide better drivability at 1,000 feet plus above sea level.

Ambient Temperature Temperature of the surrounding air. Normally, it is considered to be the temperature in the service area where testing is taking place.

Amp Acronym for ampere.

Ampere The unit for measuring electrical current.

Analog Signal A voltage signal that varies within a given range (from high to low, including all points in between).

Analog-to-Digital Converter (A/D Converter) A device that converts analog voltage signals to a digital format; located in the ECM.

Analog Volt/Ohmmeter (AVOM) A test meter used for checking voltage and resistance. Analog meters should not be used on solid state circuits.

Antifreeze A mixture added to water to lower its freezing point.

Armature The rotating component of a (1) starter or other motor. (2) generator.

Articulation Pivoting movement.

ASE Acronym for Automotive Service Excellence, a trademark of National Institute for Automotive Service Excellence.

Atmospheric Pressure The weight of the air at sea level; 14.696 pounds per square inch (psi) or 101.33 kilopascals (kPa).

Axis of Rotation The center line around which a gear or part revolves.

Battery Terminal A tapered post or threaded studs on top of the battery case for connecting the cables.

Bimetallic Two dissimilar metals joined together that have different bending characteristics when subjected to changes of temperature.

Blade Fuse A type of fuse having two flat male lugs for insertion in female box connectors.

Blower Fan A fan that pushes or blows air through a ventilation, heater, or air conditioning system.

Bobtailing A tractor running without a trailer.

Boss Outer race of a bearing.

Bottoming A condition that occurs when the teeth of one gear touch the lowest point between teeth of a mating gear.

British Thermal Unit (BTU) A measure of heat quantity equal to the amount of heat required to raise 1 pound of water 1°F.

BTU Acronym for British Thermal Unit.

CAA Acronym for Clean Air Act.

Cartridge Fuse A type of fuse having a strip of low melting point metal enclosed in a glass tube. If an excessive current flows through the circuit, the fuse element melts at the narrow portion, opening the circuit and preventing damage.

Caster The angle formed between the kingpin axis and a vertical axis as viewed from the side of the vehicle. Caster is considered positive when the top of the kingpin axis is behind the vertical axis.

Cavitation A condition caused by bubble collapse.

C-EGR Acronym for cooled exhaust gas recirculation.

CFC Acronym for chlorofluorocarbon.

Charging System A system consisting of the battery, alternator, voltage regulator, associated wiring, and the electrical loads of a vehicle. The purpose of the system is to recharge the battery whenever necessary and to provide the current required to power the electrical components.

Charging Circuit The alternator (or generator) and associated circuit used to keep the battery charged and power the vehicle's electrical system when the engine is running.

Check-Valve A valve that allows air to flow in one direction only.

Climbing A gear problem caused by excessive wear in gears, bearings, and shafts whereby the gears move sufficiently apart to cause the apex of the teeth on one gear to climb over the apex of another gear.

Clutch A device for connecting and disconnecting the engine from the transmission.

COE Acronym for cab-over-engine.

Coefficient of Friction A measurement of the amount of friction developed between two objects in physical contact when one of the objects is drawn across the other.

Coil Springs Spring steel spirals.

Conicity A term that describes the tire's tendency to roll like a cone.

Compression Applying pressure to a spring or fluid.

Compressor Mechanical device that increases pressure within a circuit.

Condensation The process by which gas (or vapor) changes to a liquid.

Conductor Any material that permits the electrical current to flow.

Coolant Liquid that circulates in an engine cooling system.

Coolant Heater A component used to aid engine starting and reduce the wear caused by cold starting.

Coolant Hydrometer A tester designed to measure coolant specific gravity and determine antifreeze protection.

Cooling System A system for circulating coolant.

Crankcase The housing within which the crankshaft rotates.

Cranking Circuit The starter circuit, including battery, relay (solenoid), ignition switch, neutral start switch (on vehicles with automatic transmission), and cables and wires.

Cycling (1) On-off action of the air conditioner compressor. (2) Repeated electrical cycling that can cause the positive plate material to break away from its grids and fall into the sediment base of the battery case.

Dampen To slow or reduce oscillations or movement.

Dampened Discs Discs that have dampening springs incorporated into the disc hub. When engine torque is transmitted to the disc, the plate rotates on the hub, compressing the springs. This action absorbs the torsional vibration caused by today's low RPM, high-torque, engines.

Data Links Circuits through which computers communicate with other electronic devices such as control panels, modules, sensors, or other computers.

Deburring To remove sharp edges from a cut.

Deflection Bending or moving to a new position as the result of an external force.

Detergent Additive An additive that helps keep metal surfaces clean and prevents deposits. These additives suspend particles of carbon and oxidized oil in the oil.

DER Acronym for Department of Environmental Resources.

Diagnostic Flow Chart A chart that provides a systematic approach to the electrical system and component trouble-shooting and repair. They are found in service manuals and are vehicle make and model specific.

Dial Caliper A measuring instrument capable of taking inside, outside, depth, and step measurements.

Digital Binary Signal A signal that has only two values; on and off.

Digital Volt/Ohmmeter (DVOM) A test meter recommended for use on solid state circuits.

Diode Semiconductor device formed by joining P-type semiconductor material with N-type semiconductor material. A diode allows current to flow in one direction, but not in the opposite direction.

DOT Acronym for Department of Transportation.

Driven Gear A gear that is driven by a drive gear, by a shaft, or by some other device.

Drive or Driving Gear A gear that drives another gear.

Driveline The propeller or driveshaft, and universal joints that link the transmission output to the axle pinion gear shaft.

Driveline Angle The alignment of the transmission output shaft, driveshaft, and rear axle pinion centerline.

Driveshaft Assembly of one or two universal joints connected to a shaft or tube; used to transmit torque from the transmission to the differential.

Drivetrain An assembly that includes all torque transmitting components from the rear of the engine to the wheels.

ECM Acronym for electronic control module.

ECU Acronym for electronic control unit.

Eddy Current Circular current produced inside a metal core in the armature of a starter motor. Eddy currents produce heat and are reduced by using a laminated core.

Electricity The movement of electrons from one location to another.

Electromotive Force (EMF) The force that moves electrons between atoms. This force is the pressure that exists between the positive and negative points. This force is measured in units called volts; charge differential.

Electronically Erasable Programmable Memory (EEP-ROM) Computer memory that enables write-to functions.

Electrons Negatively charged particles orbiting every nucleus.

EMF Acronym for electromotive force.

Engine Brake A hydraulically operated device that converts the engine into a power absorbing mechanism.

Environmental Protection Agency An agency of the United States government charged with the responsibilities of protecting the environment.

EPA Acronym for the Environmental Protection Agency.

Exhaust Brake A slide mechanism which restricts the exhaust flow, causing exhaust back-pressure to build up in the engine's cylinders. The exhaust brake actually transforms the engine into a power absorbing air compressor driven by the wheels.

False Brinelling The polishing of a surface that is not damaged.

Fatigue Failures Progressive destruction of a shaft or gear teeth usually caused by overloading.

Fault Code A code that is recorded into the computer's memory.

Federal Motor Vehicle Safety Standard (FMVSS) A federal standard that specifies that all vehicles in the United States be assigned a Vehicle Identification Number (VIN).

Fixed Value Resistor An electrical device that is designed to have only one resistance rating, which should not change, for controlling voltage.

Flammable Any material that will easily catch fire or explode.

Flare To spread gradually outward in a bell shape.

Foot-Pound An English unit of measurement for torque. One foot-pound is the torque obtained by a force of 1 pound applied to a foot long wrench handle.

Fretting A result of vibration that the bearing outer race can pick up the machining pattern.

Fusible Link A term often used for an insulated fuse link.

Fuse Link A short length of smaller gauge wire installed in a conductor, usually close to the power source.

Gear A disk-like wheel with external or internal teeth that serves to transmit or change motion.

Gear Pitch The number of teeth per given unit of pitch diameter, an important factor in gear design and operation.

Geometric Center Line A line drawn through the center points of the front and rear axle.

Ground The negatively charged side of a circuit. A ground can be a wire, the negative side of the battery, or the vehicle chassis.

Grounded Circuit A shorted circuit that causes current to return to the battery before it has reached its intended destination.

Harness and Harness Connectors The vehicle's electrical system wiring providing a convenient starting point for tracking and testing circuits.

Hazardous Materials Any substance that is flammable, explosive, or is known to produce adverse health effects in people or the environment.

Heads-Up Display (HUD) A technology used in some vehicles that superimposes data on the driver's normal field of vision. The operator can view the information, which appears to "float" just above the hood at a range near the front of a conventional tractor or truck. This allows the driver to monitor conditions such as road speed without interrupting his view of traffic.

Heater Control Valve A valve that controls the flow of coolant into the heater core from the engine.

Heat Exchanger A device used to transfer heat, such as a radiator or condenser.

Heavy-Duty Truck A truck that has a GVW of 26,001 pounds or more.

High-Resistant Circuits Circuits that have an increase in circuit resistance, with a corresponding decrease in current.

High-Strength Steel A low-alloy steel that is stronger than hot-rolled or cold-rolled sheet steels.

Hinged Pawl Switch The simplest type of switch; one that makes or breaks the current of a single conductor.

HUD Acronym for heads-up display.

Hydrometer A tester designed to measure the specific gravity of a liquid.

Inboard Toward the centerline of the vehicle.

Included Angle An angle determined by adding the camber and steering angle inclination (SAI) together on any one wheel.

In-Line Fuse A fuse that is in series with the circuit in a small plastic fuse holder, not in the fuse box or panel. It is used, when necessary, as a protection device for a portion of the circuit even though the entire circuit may be protected by a fuse in the fuse box or panel.

Installation Templates Drawings supplied by some vehicle manufacturers to allow the technician to correctly install the accessory. The templates available can be used to check clearances or to ease installation.

Insulator A material, such as rubber or glass, that offers high resistance to electron flow.

Integrated Circuit A component containing diodes, transistors, resistors, capacitors, and other electronic components mounted on a single piece of material and capable to perform numerous functions.

Jacobs Engine Brake An engine brake, named for its inventor. A hydraulically operated device that converts a power producing diesel engine into a power-absorbing retarder.

Jumper Wire A wire used to temporarily bypass a circuit or components for electrical testing. A jumper wire consists of a length of wire with an alligator clip at each end.

Jump Start The procedure used when it becomes necessary to use a boost battery to start a vehicle with a discharged battery.

Kinetic Energy Energy in motion.

Lateral Runout The wobble or side to side movement of a rotating wheel.

Laser Beam Alignment System A two- or four-wheel alignment system using wheel-mounted instruments to project a laser beam to measure toe, caster, and camber.

Linkage A system of rods and levers used to transmit motion or force.

Low-Maintenance Battery A conventionally vented, lead/acid battery, requiring normal periodic maintenance.

Magnetorque An electromagnetic clutch.

Maintenance-Free Battery A battery that does not require the addition of water during normal service life.

Maintenance Manual A publication containing routine maintenance procedures and intervals for vehicle components and systems.

Memory Steering Occurs when a steering component or bushing binds and prevents the steering gear from smoothly rotating back to center.

NATEF Acronym for National Automotive Education Foundation.

National Automotive Technicians Education Foundation (NATEF) A foundation having a program of certifying secondary and post-secondary automotive and heavy-duty truck training programs.

National Institute for Automotive Service Excellence (ASE) A nonprofit organization that has an established certification program for automotive, heavy-duty truck, auto body repair, engine machine shop technicians, and parts specialists.

NIOSH Acronym for National Institute for Occupation Safety and Health.

NLGI Acronym for National Lubricating Grease Institute.

NHTSA Acronym for National Highway Traffic Safety Administration.

NOP Acronym for Nozzel Opening Pressure. Pressure in an injector nozzle opens at inoperation. Also known as VOP.

OEM Acronym for original equipment manufacturer.

Off-road With reference to unpaved, rough, or ungraded terrain on which a vehicle will operate. Any terrain not considered part of the highway system falls into this category.

Ohm A unit of electrical resistance.

Ohm's Law Basic law of electricity stating that in any electrical circuit, current, resistance, and pressure work together in a mathematical relationship.

On-Road With reference to paved or smooth-graded surface on which a vehicle will operate; part of the public highway system.

Open Circuit An electrical circuit whose path has been interrupted or broken either accidentally (a broken wire) or intentionally (a switch turned off).

Oscillation Movement in either fore/aft or side to side direction about a pivot point.

OSHA Acronym for Occupational Safety and Health Administration.

Out of Round Eccentric.

Output Driver Electronic switch that the computer uses to control the output circuit. Output drivers are located in the output ECM.

Oval Condition that occurs when a tube is egg-shaped.

Overrunning Clutch A clutch mechanism that transmits power in one direction only.

Overspeed Governor A governor that shuts off fuel at a specific RPM.

Oxidation Inhibitor An additive used with lubricating oils to keep oil from oxidizing at high temperatures.

Parallel Circuit An electrical circuit that provides two or more paths for current flow.

Parallel Joint Type A type of driveshaft installation whereby all companion flanges and/or yokes in the complete driveline are parallel to each other with the working angles of the joints of a given shaft being equal and opposite.

Parking Brake A mechanically applied brake used to prevent a parked vehicle's movement.

Parts Requisition A form that is used to order new parts, on which the technician writes the part(s) needed along with the vehicle's VIN.

Payload The weight of the cargo carried by a truck, not including the weight of the body.

Pitting Surface irregularities resulting from corrosion.

Polarity The state, either positive or negative, of charge differential.

Pole The number of input circuits made by an electrical switch.

Pounds per Square Inch (psi) A unit of English measure for pressure.

Power A measure of work being done factored with time.

Power Flow The flow of power from the input shaft through one or more sets of gears.

Power Train Engine to the wheels in a vehicle.

Pressure The force applied to a definite area measured in pounds per square inch (psi) English or kilopascals (kPa) metric.

Pressure Differential The difference in pressure between any two points of a system or a component.

Printed Circuit Board Electronic circuit board made of thin nonconductive material onto which conductive metal has been deposited. The metal is then etched by acid, leaving lines that form conductors for the circuits on the board. A printed circuit board can hold many complex circuits.

Programmable Read-Only Memory (PROM) Electronic component that contains program information specific to vehicle model calibrations.

PROM Acronym for Programmable Read-Only Memory.

Psi Acronym for pounds per square inch.

P-type Semiconductors Positively biased semiconductors.

RAM Acronym for random access memory; main memory.

Ram Air Air forced into the engine housing or passenger compartment by the forward motion of the vehicle.

Random Access Memory (RAM) Memory used during computer operation to store temporary information. The microcomputer can write, read, and erase information from RAM; electronically retained.

RCRA Acronym for Resource Conservation and Recovery Act.

Reactivity The characteristic of a material that enables it to react violently with air, heat, water, or other materials.

Read-Only Memory (ROM) Memory used in micro-computers to store information permanently.

Recall Bulletin A bulletin that pertains to special situations that involve service work or replacement of components in connection with a recall notice.

Reference Voltage The voltage supplied to a sensor by the computer, which acts as baseline voltage; modified by the sensor to act as an input signal; usually 5 VDC.

Relay An electric switch that allows a small current to control a much larger one. It consists of a control circuit and a power circuit.

Reserve Capacity Rating The ability of a battery to sustain a minimum vehicle electrical load in the event of a charging system failure.

Resistance Opposition to current flow in an electrical circuit.

Resource Conservation and Recovery Act (RCRA) Law that states that after using hazardous material, it must be properly stored until an approved hazardous waste hauler arrives to take it to a disposal site.

Revolutions per Minute (RPM) The number of complete turns a shaft turns in one minute.

Right to Know Law A law passed by the federal government and administered by the Occupational Safety and Health Administration (OSHA) that requires any company that uses or produces hazardous chemicals or substances to inform its employees, customers, and vendors of any potential hazards that may exist in the workplace as a result of using the products.

Ring Gear The gear around the edge of a flywheel.

ROM Acronym for read only memory.

Rotary Oil Flow A condition caused by the centrifugal force applied to the fluid as the converter rotates around its axis.

Rotation A term used to describe a gear, shaft, or other device when it is turning.

RPM Acronym for revolutions per minute.

Rotor Rotating part of the alternator that provides the magnetic fields necessary to create a current flow. The rotating member of an assembly.

Runout Deviation or wobble of a shaft or wheel as it rotates. Measured with a dial indicator.

Semiconductor Solid state material used in diodes and transistors.

Sensing Voltage The voltage that allows the regulator to sense and monitor the battery voltage level.

Sensor An electronic device used to monitor conditions for computer control requirements; an input circuit device.

Series Circuit A circuit connected to a voltage source with only one path for electron flow.

Series/Parallel Circuit A circuit designed so that both series and parallel conditions exist within the same circuit.

Service Bulletin Publication that provides the latest service tips, field repairs, product improvements, and related information of benefit to service personnel.

Service Manual A manual, published by the manufacturer, that contains service and repair information for all vehicle systems and components.

Short Circuit An undesirable connection between two worn or damaged wires. The short occurs when the insulation is worn between two adjacent wires and the metal in each wire contacts the other, or when wires are damaged or pinched.

Single-Axle Suspension A suspension with one axle.

Single Reduction Axle Any axle assembly that employs only one gear reduction through its differential carrier assembly.

Solenoid An electromagnet that is used to conduct electrical energy in mechanical movement.

Solid Wire Single-strand conductor.

Solvent Substance which dissolves other substances.

Spade Fuse Term used for a blade fuse.

Spalling Surface fatigue that occurs when chips, scales, or flakes of metal break off.

Specialty Service Shop A shop that specializes in areas such as engine rebuilding, transmission/axle overhauling, brake, air conditioning/heating repairs, and electrical/electronic work.

Specific Gravity The ratio of a liquid's mass to that of an equal volume of distilled water.

Spontaneous Combustion A reaction in which a combustible material self-ignites.

Stall Test Test performed when there is a malfunction in the vehicle's power package (engine and transmission), to determine which of the components is at fault.

Starter Circuit The circuit that carries the high current flow and supplies power for engine cranking.

Starter Motor Device that converts electrical energy from the battery into mechanical energy for cranking.

Starter Safety Switch A switch that prevents vehicles with automatic transmissions from being started in gear.

Static Balance Balance at rest, or still balance.

Stepped Resistor A resistor designed to have two or more fixed values, available by connecting wires to one of several taps.

Storage Battery A battery to provide a source of direct current electricity for both the electrical and electronic systems.

Stranded Wire Wire that is made up of a number of small solid wires, generally twisted together, to form a single conductor.

Sulfation Condition that occurs when sulfate is allowed to remain on the battery plates for a long time, causing two problems: (1) It lowers the specific gravity levels, increasing the danger of freezing at low temperatures. (2) In cold weather, a sulfated battery may not have the reserve power needed to crank the engine.

Swage To reduce or taper.

Switch Device used to control on/off and direct the flow of current in a circuit. A switch can be under the control of the driver or can be self-operating through a condition of the circuit, the vehicle, or the environment.

Tachometer Instrument that indicates shaft rotating speeds.

Throw (1) Offset of a crankshaft. (2) Number of output circuits of a switch.

Thrust Angle An angle formed by the intersection of the vehicle thrust line and the geometric centerline.

Time Guide Used for computing compensation payable by the truck manufacturer for repairs or service work to vehicles under warranty.

Timing The phasing of events to produce action such as ignition.

Torque Twisting force.

Torque Converter A device, similar to a fluid coupling, that transfers engine torque to the transmission input shaft and can multiply engine torque.

Toxicity A statement of how poisonous a substance is.

Tractor A motor vehicle that has a fifth wheel and is used for pulling a semitrailer.

Transistor Electronic device produced by joining three sections of semiconductor materials. Used as a switching device.

Tree Diagnosis Chart Chart used to provide a logical sequence for what should be inspected or tested when troubleshooting a repair problem.

Vacuum Pressure values below atmospheric pressure.

Vehicle Retarder An engine or driveline brake.

VIN Acronym for Vehicle Identification Number.

Viscosity Resistance to flow or fluid sheer.

VOP Acronym for Valve Opening Pressure. Caterpillar term for NOP.

Volt The unit of electromotive force.

Voltage-Generating Sensors Devices which produce their own voltage signal.

Voltage Limiter Device that provides protection by limiting voltage to the instrument panel gauges to approximately 5 volts.

Voltage Regulator Device that controls the current produced by the alternator and thus the voltage level in the charging circuit.

Watt Measure of electrical power.

Watt's Law A law of electricity used to calculate the power consumed in electrical circuit, expressed in watts. It states that power equals voltage multiplied by current.

Wheel Tramp The term used when the tire hops up and down as it rotates.

Wheel Shimmy The side-to-side shaking that occurs when a tire is not properly balanced.

Windings (1) The three bundles of wires in the stator. (2) Coil of wire in a relay or similar device.

Work (1) Forcing a current through a resistance. (2) The product of a force.

Yield Strength The highest stress a material can stand without permanent deformation or damage, expressed in pounds per square inch (psi).

Notes

Notes